AF352702

War of Supply

War of Supply

World War II Allied Logistics in the Mediterranean

David D. Dworak

UNIVERSITY PRESS OF KENTUCKY

Scholarly publisher for the Commonwealth,
serving Bellarmine University, Berea College, Centre
College of Kentucky, Eastern Kentucky University,
The Filson Historical Society, Georgetown College,
Kentucky Historical Society, Kentucky State University,
Morehead State University, Murray State University,
Northern Kentucky University, Spalding University,
Transylvania University, University of Kentucky,
University of Louisville, and Western
Kentucky University.

Editorial and Sales Offices: The University Press of Kentucky
663 South Limestone Street, Lexington, Kentucky 40508-4008
www.kentuckypress.com

Library of Congress Cataloging-in-Publication Data

Names: Dworak, David D. (David Dean), 1962– author.
Title: War of supply : World War II allied logistics in the Mediterranean /
    David D. Dworak.
Description: Lexington, Kentucky : The University Press of Kentucky, [2021] |
    Series: Foreign military studies | Includes bibliographical references and index.
Identifiers: LCCN 2021049429 | ISBN 9780813183770 (hardcover) |
    ISBN 9780813183794 (pdf) | ISBN 9780813183800 (epub)
Subjects: LCSH: World War, 1939–1945—Campaigns—Mediterranean Region. |
    World War, 1939–1945—Logistics—Mediterranean Region.
Classification: LCC D763.M47 D86 2021 | DDC 940.54/23—dc23/eng/20211110

This book is printed on acid-free paper meeting
the requirements of the American National Standard
for Permanence in Paper for Printed Library Materials.

Manufactured in the United States of America

Member of the Association of
University Presses

*Dedicated to the thousands of men and women
in the Mediterranean Service Forces
who built the foundation upon which victory rests*

Contents

Figures

Tables

Maps

Preface

A Look behind the Curtain

Imagine that the modern battlefield is like a great stage stretching across the land-scape in some remote part of the world. On this stage we see the actors: generals devising strategy and plans; brigades and divisions maneuvering to engage the enemy; fleets of gray-hulled ships launching volleys of shells. In the sky, aircraft compete to gain an advantage over the adversary or attack strategic and tactical targets on the ground. Histories, movies, and documentaries help re-create these events, but they often tell only part of the story. How did the combatants arrive on the battlefield, far from home? How did they get the fuel, munitions, food, repair parts, clothing, water, and equipment needed to engage in great-power conflict? Who built the airfields and cleared the ports, and how did they do it? What were the support structures and doctrine that ensured the right things were in the right place at the right time? How did the combatants synchronize support requirements with available resources? In short, we see the actors in front of us, but what about the story of everything that happened to make the production possible? Like a play, what were the decisions, actions, challenges, and events that took place behind the curtain, setting the foundation for success?

This story is about the challenges of establishing and maintaining a theater of war in the modern age, specifically looking at the Allied experience in the Mediterranean from 1942 to 1945. Military logistics and administrative support rarely receive much attention, but history frequently demonstrates they are often of critical importance in the outcome of the battle, operation, or campaign. Like a stage production, logistics and support efforts set the foundation for the event, often determining the realm of what might be possible. An effective support system provides combat power to commanders, enables operations, and sets conditions for follow-on operations. An ineffective logistics system limits options and can prevent commanders from seizing and exploiting opportunities. Perhaps there is a reason doctrine refers to the geographic region that incorporates both the combat and support zones as a "theater of war." Like a theater, the war zone needs actors (combatants), but it also needs myriad others to help make the production possible. This is the story of the thousands of men and women who worked behind the scenes supporting Allied efforts in the Mediterranean in World War II, setting the foundation for victory in the Mediterranean and Europe. The insights and lessons they gained through trial and error have relevancy even today.

Abbreviations

AFHQ	Allied Force Headquarters
BS	Base Section
COMZONE	communications zone
COSSAC	chief of staff to the supreme Allied commander
D-Day	first day of an operation
G3	operations officer
G4	logistics and support officer
LCT	Landing Craft, Tank
LCVP	Landing Craft, Vehicle Personnel
LST	Landing Ship, Tank
MTOUSA	Mediterranean Theater of Operations, United States Army
NATOUSA	North African Theater of Operations, United States Army
POW	prisoner of war
SHAEF	Supreme Headquarters Allied Expeditionary Force
SOS	Services of Supply
SUP	single-unit pack
TUP	twin-unit pack

Introduction

> The handling of the supply problem is of no less importance than operational
> and tactical command.
> —Field Marshal Alfred Kesselring

In November 1942, the United States initiated its first major offensive actions against forces of the Axis European nations. For the next two and a half years, the United States and its allies engaged in a series of battles across the Mediterranean that had a profound impact on the outcome of World War II. However, victory in the Mediterranean was the result of more than good generalship, an effective strategy, unit engagements, or even luck. Success for either side also depended on the ability to support military forces thousands of miles away from their home bases.

World War II was one of matériel. The armies, navies, and air forces of the mid-twentieth century were growing ever dependent on the machines and technology of modern war. Tanks, trucks, mobile artillery, aircraft, and ships of war needed vast amounts of fuel, munitions, and repair parts. Most American divisions were motorized or mechanized, meaning that they moved by some type of vehicle and could cover hundreds of miles in a week. Fighter and bomber aircraft developed ranges and capabilities that were unheard of in World War I. However, the militaries of World War II also needed other more mundane but no less important types of supplies and support to perform on the battlefield. Men needed food, medical supplies, tents, water, uniforms, and individual equipment. Units needed trucks, Jeeps, typewriters, binoculars, and a host of other gear. Along with this equipment came the need for medical, engineer, supply, and transportation units as well as for repair parts and mechanics.

This war of matériel was not just for the Allies; Axis forces also felt the impact of industrial modernization. In 1946, Field Marshal Albert Kesselring, commander of German forces in the Mediterranean and later the German commander in chief–West assessed that one of the major reasons behind Allied victory was a commonality of repair parts among the different forces combined with an effective supply system—things the German military often lacked. In Kesselring's opinion, "the handling of the supply problem is of no less importance than operational and tactical command."[1]

War in the 1940s was more than just fighting. Indeed, war was also about producing the units, equipment, and supplies needed for combat and then shipping them halfway across the world to a remote beach or port. Once at the point of debarkation, an administrative organization had to offload, organize, and transport everything to the front. Support units had to build installations and airfields, establish factories for the assembly of vehicles, and create an administrative bureaucracy to manage the entire administrative effort to support a theater of war. Added to all this, the Allied militaries had to provide food and medical care for civil populations, outfit allies who could not support themselves, as well as house and care for tens of thousands of prisoners of war. The divisions on the front lines were important, but increasingly important were the support units operating in the rear areas that handled all of the administrative functions. Service forces such as quartermaster, transportation, ordnance, Medical Corps, Women's Army Corps, and engineer units set the conditions that enabled the combat units to fight and win.

The Allied fight for the Mediterranean weaved its way through a series of campaigns—North Africa, Sicily, Italy, and southern France. The invasion of North Africa was the first major combined expeditionary landing of US and British forces in support of the European theater. Sicily served as a test bed for the Allies to correct problems seen in North Africa and served to clear Mediterranean shipping lanes. The first major Allied invasion of the European continent was of Italy; the invasion of southern France provided a much-needed secondary avenue of approach into Germany. Behind all these campaigns and battles were another series of engagements both within the Allied coalition and between the US military services—clashes over command, relevancy, priority, and resourcing. All sides had to face the tyranny of physics, time, and distance. These nonbattle skirmishes played an important part in Allied operations as forces struggled to refine doctrine, establish workable systems of command and control, and learn to work in an environment where requirements often exceeded capability. As the Allies discovered in North Africa, contemporary warfare was as much about what occurred leading up to and behind the combat zone as about the battles that took place within it. Logistics alone may not win the war, but absent or ineffective logistics support *can* lose it.

The Mediterranean campaign shows that almost everything concerning support was a struggle and nothing was easy. The need to get the right mix of combat forces and sustainment onto the beach at the right time was critical because it would determine whether a force could land and stay on the beach or be pushed off. This is a tale of massive requirements, scarce resources, short planning timelines, innovation, competing demands, changing proprieties, and organizational improvement. Through trial and error, Allied forces learned to support large invasions, thus setting the conditions for the later assault into northwestern Europe.

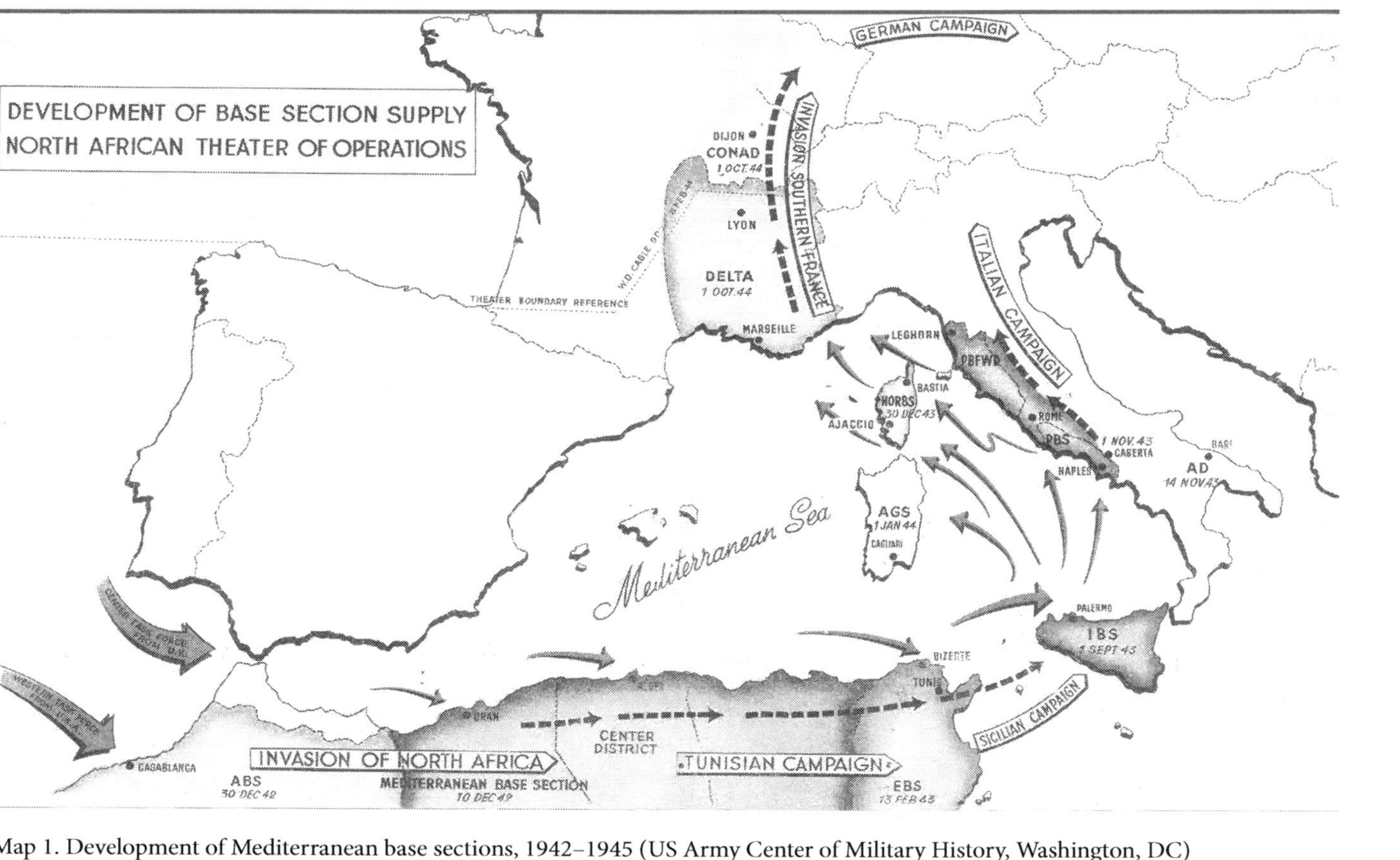

Map 1. Development of Mediterranean base sections, 1942–1945 (US Army Center of Military History, Washington, DC)

Table 1. Establishment of Base Sections, 1942–1945

US Base Section (BS)	Established	Location	Commander(s)
Mediterranean BS	December 1942	Oran	Brig. Gen. Tom Larkin Col. Edmond Leavey
Atlantic BS	December 1942	Casablanca	Brig. Gen. Arthur Wilson Col. Francis Oxx Col. John Ratay
Eastern BS	February 1943	Constantine	Brig. Gen. Arthur Pence Col. A. B. Conard
Island BS	September 1943	Sicily	Col. Robert Sears
Peninsular BS	November 1943	Naples	Brig. Gen. Arthur Pence Brig. Gen. Arthur Wilson Brig. Gen. Francis Oxx
Northern BS	December 1944	Corsica	Col. John Ratay
Coastal (later Continental) BS	July 1944	Marseilles	Maj. Gen. Arthur Wilson
Delta BS	October 1944	Marseilles	Brig. Gen. John Ratay

Part I

North Africa

Novice

1

Operation Torch

The Invasion of North Africa

> I don't know anything about logistics. You keep me out of trouble.
> —Major General George S. Patton Jr.

The debate over the balance between combat forces and service forces is an enduring part of modern warfare. Too many service units might mean that a military force lacks the combat power needed to defeat an adversary. Conversely, too few service forces can lead to a lack of combat effectiveness on the battlefield due to a lack of supply, high maintenance losses, or insufficient transportation. Getting the balance right is part art, part science and is extremely difficult to achieve when faced with limited resources and a complex environment.

In 1918, 34 percent of the American Expeditionary Force was in the Services of Supply (SOS) branch. Although sizable, this proportion proved inadequate, resulting in a near collapse of the US support system just prior to the armistice at the end of World War I. The interwar years did little to stress the need for support troops; in early 1942, service forces composed less than 12 percent of the US Army. As the American army entered World War II, it was not prepared to support a type of warfare that involved large numbers of mechanized and air forces capable of moving quickly across vast distances. The lack of support forces would have significant ramifications for Allied operations in North Africa.[1]

The army was not the only service experiencing growing pains within its service forces at the beginning of the war; the navy underwent challenges as well. During the interwar years, the navy elected to put its limited funds toward the building of capital ships and to improvise logistics only as the situation required. In 1942, few in the US military had any real idea of what might be required to support large fleets, air forces, and field armies deployed halfway across the globe in several theaters. Like the German military, the US Navy based logistic plans on the assumption of a short war and quick victory.[2]

Besides neglecting to make any large-scale investments in logistics shipping prior to 1942, the US Navy also failed to build an assault shipping capacity that could support large-scale invasions. Despite the American desire to invade France

"

early on in the war, there simply was an insufficient quantity of assault craft and merchant shipping to land and sustain the landing force.[3] This meant that the Allies had to look for opportunities and develop strategy that could attack the Axis forces on the peripherals without risking a direct assault on the European continent.

Prelude: 1942

Considered in total, the administrative situation leading into the summer of 1942 was not promising. Both the US Army and the US Navy had neglected to build or train a sizeable logistics capability during the interwar years. By 1942, American ports were shipping matériel faster than army service forces in England could receive it. Great Britain had been fighting Germany for two years and was expending all of its effort in supporting its own forces. The Allies had elected to modernize their forces, but equipment was in short supply, and industry was working at full capacity to meet the increased demand. The United States was building up its military presence within the British Isles, but at the same time European SOS branches were working to develop functioning organizations while also defining their responsibilities. There was a great deal of activity under way, but it was not as effective or as efficient as it could have otherwise been.[4]

To compound the Allied problem of building up military forces, the Battle of the Atlantic was ongoing throughout 1942. In that year alone, the United States lost 1,902,418 tons of merchant shipping, and the British lost 3,171,942 tons. Put in perspective, US merchant shipping losses in 1942 alone equaled *ten times* the amount of American shipping lost throughout World War I.[5]

Russia also added to the demand for Allied shipping and matériel. To keep Russia in the war, the United States shipped tremendous amounts of equipment to Russia through the Persian Gulf. President Franklin D. Roosevelt promised to ship more than 1.5 million tons of supplies to Russia by June 1942—just part of the 4,159,117 tons of Allied Russia-aid cargo that the United States shipped between November 1941 and May 1945. Not all that was promised was delivered; in addition, the shipments to allies meant that there was less equipment, less available merchant shipping, and fewer service units available for use in US combat operations. These unexpected requirements made strategic sense but ultimately resulted in an increased demand for service units. As the war effort expanded in 1942, commanders around the globe discovered theaters needed a sizeable number of service forces to function and that there just were not enough of these types of units in the inventory to satisfy demand.[6]

The German Approach to Support

Whereas the United States pursued a mechanized approach to the modernization of its military forces during the interwar years, Germany took a slightly different direction. Although a considerable amount of historical attention has focused on Adolf Hitler's armored forces and their blitzkrieg tactics, a significant percentage of the German army relied on horse-drawn vehicles throughout the war, especially in German logistics units. Only a few German units were motorized, typically the first-line mechanized divisions, and only the German armored divisions with their accompanying motorized infantry divisions were totally free of horse-drawn vehicles.[7]

Of the 102 German divisions that existed in 1939, only 14 were fully mechanized or motorized. In addition, the Germans had just three motor transport regiments in the entire army, with a combined cargo capacity of 19,500 tons. In comparison, for the invasion of northwestern Europe in 1944, the Allied forces landed a motor transport capability of 69,000 tons. German industry worked to field additional transport units during the war, but it was never able to divest the military of its dependence on horse transport. No horses were used in North Africa due to the climate, which helped to contribute to a false notion that has endured in the historical narrative of a German army characterized by large mechanized formations, lightning-rapid movements, and operational agility.[8]

The Germans lacked widespread motor transport for several reasons. First, Germany lacked natural resources, especially iron and oil. Second, there was no government system in place as of 1939 that could effectively convert civilian automobile production to military transport. Last, as war broke out, Germany had the lowest ratio of automobiles to population of any European nation, except for Italy. This meant not only that Germany was unable to produce the quantities of vehicles needed to support modern warfare but also that the vehicles it produced were often more suitable for civilian use than military use, and there was little standardization across fleets.[9]

The German dependence on horse-drawn vehicles meant more than a loss of cargo capability; it also meant an increased requirement for supplies and support. Although horses did not require gas or oil to keep moving, they did need fodder and veterinary support. The requirement for fodder placed a sizeable demand on the German transportation system because fodder is relatively bulky compared to its weight. A first-line infantry division in 1939 had to feed an average of 4,800 horses a day. Each second- and third-line division maintained more than 6,000 horses, and each horse ate between twelve and twenty pounds of fodder a day, meaning that an average division's fodder requirements were approximately 43 tons *per day*.[10]

Germany ultimately fielded almost twice the number of horses during World War II (2,700,000) as it had in World War I (1,400,000). The Allied forces, in comparison, had virtually no horse-drawn transport in their armies during World War II and relied on pack mules only for use in terrain that was unsuitable for wheeled vehicles.[11]

The German reliance on horse transport makes some sense when considering the High Command's (Oberkommando der Wehrmacht's) approach to war. During the early years of the war, Germany was largely able to operate on internal lines of communication over good roads and, more importantly, with an established and sizeable rail network. German logisticians estimated that one 200-mile stretch of double track railway equaled the capacity of 1,600 trucks. As such, continental Germany could afford to assume risk by not motorizing its field armies and support organizations.[12]

However, the war developed into a conflict of attrition involving great distances and distant theaters of operations. Infantry officers had to learn the limitations of working with tanks, but armor officers knew little about working with formations that relied on large numbers of horses for transport. German maintainers had to deal with fleets of vehicles that had little standardization. Compounding the situation was a lack of petroleum-refinement facilities. During the war, the Allied and neutral powers accounted for 90 percent of the world's refinement capability, whereas Germany and Italy accounted for a mere 4 percent. The German approach to support of the country's military forces would prove to be a deciding factor with strategic consequences.[13]

Not Exactly a Model of Planning

The US military would discover that war in northern Africa was totally unlike anything it had ever experienced. Unlike continental Europe, North Africa lacked significant civilian infrastructure. Roads were unimproved and limited, tending to follow the coastline. Rail lines were extremely limited and tended to consist of a single line, which also followed the coastline. To complicate matters, rail gauges changed from one nation to the next. The desert environment was harsh on both men and machines.

Perhaps most significant was the great distance between ports and the armies. The distance between Casablanca and Tunis was 1,028 miles. The eventual German retreat from El Alamein to Tunis would cover 1,400 miles. The entire area of operations covered more than a million square miles. To operate across such distances, the combatants, Axis and Allied alike, needed to make a mental adjustment to the scale of the battlefield and to the sheer volume of supplies, equipment, and service forces needed to sustain combat operations across these extended lines of communication—circumstances not experienced in World War I. In such conditions, the theater logistics effort determined which strategies were viable and could easily determine the outcome of a battle before the first shot fired. The Allies did not realize it then, but the invasion of North Africa was only the first of what would become a series of campaigns across the Mediterranean—campaigns that would eventually stretch from Africa to Sicily, Italy, and southern France. It was the first step toward an eventual victory against Germany, a victory that the Allies could not have achieved as quickly as they did without the experiences gained in the Mediterranean.

On July 25, 1942, President Franklin D. Roosevelt decided to go ahead with the North African invasion, so the decision was made only three months before the anticipated landing date. Logistics planning for Operation Torch began shortly thereafter, progressing at a frantic pace through August. The US War Department's SOS branch called in the chiefs of services, and work began on plans to support a total force of 100,000 to 125,000 men. The planning for Torch was hurried—one report on the planning effort was especially candid: "Neither the British nor the American General Staffs would care to offer Torch as a model of planning."[14]

The operational objectives for Torch were twofold. First, execute the landings, secure the ports, and establish lines of communication. Second, trap and destroy German general Erwin Rommel's forces between the British First and Eighth Armies. This had to be done in an austere environment with little infrastructure before the winter rains came and before the Axis could significantly reinforce North Africa.[15]

Although the Mediterranean theater was a joint venture between American and British forces, each nation established independent organizations for support of its effective parts of the battlefield. Due to a variety of reasons—not the least of which was the lack of compatibility between supply systems—each nation supported the forces under its own command. Thus, two different, parallel national support systems developed and operated simultaneously within the Mediterranean. Operation Torch had two major phases: simultaneous landings in French Morocco and Algeria conducted by three task forces, followed by an overland advance across Algeria, then east and into Tunisia. Commanders hoped to land in North Africa, convince the local French forces to join the Allied cause, and then join the fight against Italian and German forces operating in Tunisia.

The Western Task Force was American, comprising the US 3rd and 9th Infantry Divisions, the 2nd Armored Division, and a number of smaller supporting units. The force totaled approximately 44,760 men and was under the command of Major General George S. Patton Jr. The Western Task Force had the mission of making amphibious landings in western French Morocco, occupy ports and airfields, and then build up sufficient strength to occupy Spanish Morocco if needed.[16]

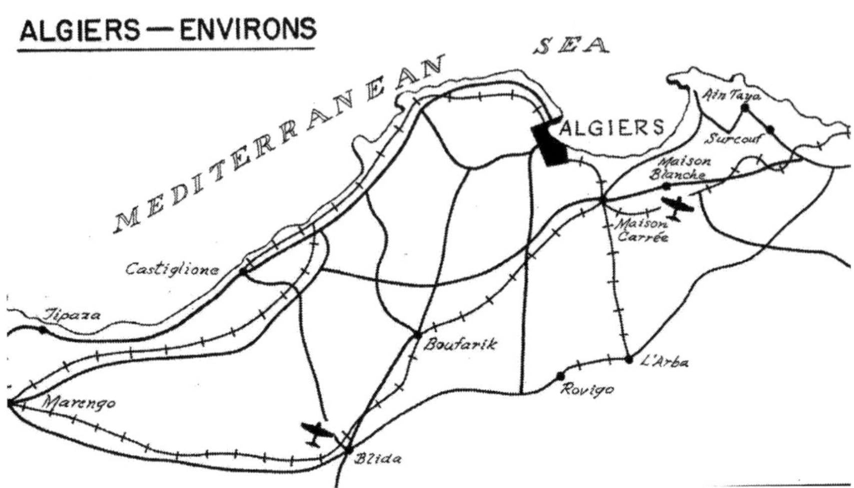

Map 2. Algerian airfields, from a Twelfth Air Force report of operations, "The AAF in North Africa" (1943?) (Center for Air Force History, Maxwell Air Force Base, AL)

The Center Task Force consisted of the US 1st Armored Division, 1st Infantry Division, and 1st Ranger Battalion. Overall, this task force contained approximately 37,000 Americans and 3,600 British and was commanded by Major General Lloyd Fredendall. The Center Task Force was to land on several beaches around Oran, Algeria, in order to seize the port and airfields and, like the Western Task Force, to be ready to occupy Spanish Morocco.[17]

The Eastern Task Force was predominately a British force, including the British 78th Division and 6th Commando as well as the 168th Combat Team of the US 34th Infantry Division and the 39th Combat Team of the US 9th Infantry Division. In total, the Eastern Task Force contained some 45,000 British and 10,000 US soldiers. Due to a history of political animosity between the French and British, a US officer, Major General Charles Ryder, commander of the 34th Infantry Division, would initially command the Eastern Task Force during the landings, and then the command would pass at a "suitable time" to the British First Army commander, Lieutenant General Kenneth Anderson, for the assault on Tunisia. Upon landing, the Eastern Task Force was to seize Algiers and the airfields at Blida and Maison Blanche. The British First Army would then move east and seize Tunisia as soon as reasonably possible, a distance of up to 400 miles.[18]

Each task force was responsible for organizing its support plan and support forces. Brigadier General Arthur Wilson commanded the Western Task Force support forces, and Brigadier General Thomas Larkin commanded the Center Task

Force SOS. Planners expected that Wilson and Larkin would operate independently under control of their respective task forces and that there *could* be a future possibility that the two task force support elements would merge and the resulting theater SOS would fall directly under Allied Force Headquarters (AFHQ). Because the Eastern Task Force had no US supply organization, support for US forces in the East would come from British supply units. In the months leading to the sailing of the convoys, AFHQ logisticians focused on the immediate tasks required to embark the force and support the landings, leaving details of a subsequent theater support organization to be determined as the operation progressed.[19]

US and British officers conducted much of the planning for Torch in London, which made coordination with the US War Department SOS challenging. There was not a lot of information on North Africa available to planners, so that much of the data concerning ports, roads, and other infrastructure came from past issues of *National Geographic* magazine. Although Casablanca was far from Tunisia, it was an important objective because it was beyond the reach of German fighters and was at the end of a single rail line that ran east through the Atlas Mountains to Oran, Algiers, and terminated in Tunisia. This single railway provided a direct route to the Tunisian objectives that would secure North Africa. The railroad had only a limited capacity, but it did provide a capability that replaced some truck transport requirements.[20]

The lack of a ground tactical plan initially hampered logistics-planning efforts because there was no definitive sailing date for the convoys and no information on composition of the assault echelons. Some of this information did not arrive for a month or more after planners had requested it. By August 20, 1942, planners finally had a troop list, and work was progressing to equip deploying units and to assemble the vast quantity of supplies and replacement equipment needed to sustain the force once it landed in North Africa.[21]

Even though work had begun to outfit the deploying units, as late as the middle of September there still was no central plan of operations approved by both the United States and Great Britain—less than forty-eight days prior to the sail date of the first convoy. Logistics planners made educated guesses on the types and quantities of needed matériel. In all cases, requisitions for supplies and equipment had to take production and shipment times into account. This meant that units sometimes ordered supplies with only a general idea of the overall requirements in order for the matériel to be on hand and loaded prior to the projected sail dates. The physics of time and distance could not wait for the detailed combat plans to be fully developed.[22]

The limited planning time resulted in confusion and hurried plans, which led to late revisions and changes in forces. This effort was essentially an exercise in trial and error for planning and executing a large-scale amphibious attack that involved the combined forces of several nations—a logistician's nightmare. Physics and time regulated the logistics system, but combat commanders and their planners were working in another realm with little regard for the logistical implications of their late decisions and changes in units.[23]

By September, the Allies had agreed to have most of the service units for the Center and Western Task Forces deploy from the United States. Service units already in Great Britain needed to prepare for an eventual direct assault on the European continent. A large percentage of equipment for the Center Task Force also had to come from the United States even though the War Department had already shipped a great deal of matériel to Great Britain. The British had enacted a policy of splitting up convoys to take advantage of available port capacity, and, as a result, accountability for much of the equipment and supplies sent to the England in 1942 was initially lost. Supply officers generally knew what had arrived in the United Kingdom but knew little regarding the specific whereabouts of needed items. As the SOS struggled to locate stocks, the requirements list grew. By September 8, the list of missing supplies and equipment amounted to 244,000 tons of stocks that would need to be replaced and shipped from the United States—a staggering amount. In addition, much of the artillery ammunition shipped to England was unusable due to damage incurred in transport. These problems were signs of an amateurish army with an unskilled force and immature support structures.[24]

As the sail date for the convoys approached, the impact of preparing logistics plans prior to tactical plans became more evident. There seemed to be little coordination between the two groups of planners: the logisticians worked to get supplies to the port, while task force planners prepared ground concepts that produced an entirely different set of requirements. On the one hand, support planners complained that Patton's staff seemed to be creating plans with little regard for the physical restraints of port capacity and convoy limitations. On the other hand, Patton's staff worked in several different locations and developed its plans in parallel with other tactical-planning efforts. Patton provided little guidance, telling his logistics and support officer (G4), Colonel Walter Muller, "I don't know anything about logistics. You keep me out of trouble." Given the short time allowed for planning, the different groups that were conducting planning, and the lack of involvement by some senior commanders, it is not surprising that the combat and support plans for Operation Torch were neither well synchronized nor entirely supportable.[25]

Similarly, loading operations for Torch proved to be chaotic and inefficient. During August and September, each of the task forces planned heavily for supplies, anticipating the need for a level of stocks that could support operations for between sixty and ninety days. Accordingly, supply officers provided requisitions to the depots, which then worked to fill as many of the demands as possible. The ports quickly became overwhelmed by the arriving shipments, and quartermasters worked to link up the arriving supplies with the right ships. In a sweeping about-face, AFHQ issued a directive on October 16, 1942, that reduced the ration load to only forty-five days in order to free up space for additional vehicles and equipment. This change required an almost complete unloading of *all* rations from the ships to achieve a balance of different food items for the forty-five-day level of supply, which then had to be reloaded.[26]

As the Allied forces, along with the associated supplies and equipment, assembled at ports along the US East Coast and in Great Britain in the fall of 1942, Lieutenant General Dwight D. Eisenhower and Major General Mark W. Clark made a decision that would later have a profound impact on the North African landings and on the greater campaign. Because merchant shipping space was at the time limited due to a lack of navy escort vessels, Clark made the call to significantly reduce the amounts of vehicles and supplies but maintain the same number of combat forces. Half of the cargo trucks were to be left on the piers. Eisenhower endorsed this decision, in part because he believed that the threat of a German advance into Spain was possible. He and Clark did not think that the assault forces, especially those landing in the West, would have to move quickly, so the vehicles could arrive on subsequent convoys. This assumption would prove to have serious consequences in terms of both the sustainment and mobility of the Allied force—consequences that would become apparent as soon as the assault force came ashore on the North African coast.[27]

Using hindsight, one can easily deride Eisenhower's decision to reduce the number of cargo vehicles, but there were compelling reasons behind the decision. The Allies simply did not know how strongly the Axis powers or Vichy French would react to the invasion. Some analysts believed that Germany might move into Spain and take over airfields, thus placing Axis aircraft within range of the landing beaches. There was also a possibility of attacks from Spanish Morocco, a Spanish colony located immediately to the north of French Morocco and straddling the landing sites of the Western and Center Task Forces. In short, the Allies had little faith in Spanish neutrality, and Eisenhower had to be prepared to defend his western flank, which led to the decision to load more combat units onto the assault convoy at the expense of the service forces.

Eisenhower approved the final plan for supply for Torch on October 27, 1942, four days *after* the initial convoys had departed. The plan called for the task forces to send all requisitions through AFHQ for approval. Following that, AFHQ would send the requisitions back to the European Theater of Operations in the United Kingdom to be filled from existing stocks. Any requisitions that Europe could not fill made their way to the US War Department SOS. Eisenhower's goal was to assemble a ninety-day level of supplies within the Torch area of operations and a sixty-day level of reserves in Great Britain. Logisticians hoped to take advantage of a shorter distance between the United Kingdom and North Africa; however, the time from the initiation of a requisition to the receipt of supplies was still weeks if not months.[28]

Ultimately, the US War Department deemed the proposed stockage levels unsupportable, so AFHQ again revised the plan on December 4, 1942. The new order directed a forty-five-day level of stock within North Africa for all supplies (ten for ammunition) and a thirty-day level of reserve stock for the United Kingdom. Supplies for the Western Task Force were to come directly from the United States. Most supplies for the Center Task Force would likewise come directly from the United States, except for those supplies that could fit onto allocated convoys coming from the United Kingdom. Support for the US contingent of the Eastern Task Force (one division) was to come from the United Kingdom, with British forces providing common items, such as food and fuel, to the US forces within the task force. This division of sustainment effort meant that although the majority of supplies would be coming from the United States, the theater would still have to manage several pipelines of matériel feeding into the Mediterranean theater, which would create additional challenges for theater logisticians.[29]

Allied planners envisioned resupply for the force occurring in three phases: in the initial assault phase, support units automatically sent supplies to North Africa from England and the United States. Following initial combat operations, the theater would transition into a "normal" phase, whereby supply would be semiautomatic, meaning that common items such as food and fuel would be shipped from the zone of the interior (the United States) based on unit situation reports, while items requiring more control, such as ammunition, would require specific requisitions. The final phase was to begin when the theater was fully established and communications were functioning. In this phase, the War Department would ship supplies only in response to specific requisitions. Each task force was responsible for its own administrative sustainment up until about D+40—that is, forty days after D-Day, the first day of the operation. After that time, AFHQ would establish itself in the theater and designate a combat zone and a rear area known as the "communications zone."[30]

By the third week of October, the planning was complete, the ships were loaded, and the convoys began leaving the ports and heading for North Africa. Eisenhower was worried about many things: whether German submarines would find the convoys, to what level the French would defend the beaches, and whether the Allies would have to defend against an incursion from Spanish Morocco. None of these obstacles proved to be serious. What the commander in chief should have been worried about instead was the sustainability of the force once it was ashore and whether the task forces had the capability needed to take advantage of any opportunity to seize Tunisia before the Germans had a chance to move in reinforcements. Given the information available to the Allies at the time, Eisenhower's concerns were understandable, although the landings would show that the Allies had not put the right focus on the strategic objective: the early capture of Tunisia.

More Room for Improvement: The Assault

The Allied landings started in the early morning hours of November 8, 1942, across western French Morocco (Safi, Casablanca, Fedala, Mehdia, and Port Lyautey) as well as in Algeria (Oran, Arzew, and Algiers). Each of the three task forces had a number of assigned beaches and sea/air ports within its respective area. Just as Operation Torch was not a good model of US or British planning, the landings and subsequent offloading of ships were similarly not a good model of an amphibious assault. The problems experienced during the landings negatively affected not just the buildup of combat forces ashore but also the sustainment of those forces and AFHQ's ability to move into Tunisia.

There were many reasons behind the problems experienced during the landings. Some are attributable to incomplete planning, poor assumptions, and the geography of the beaches, whereas other problems arose due to the lack of a capable beach organization. Perhaps the main reason behind the majority of the problems was the lack of training and experience across the forces and at all levels of command. Of all the different types of military operations conducted during the course of the war, those involving amphibious or airborne operations were the most complex and difficult to control.

Problems started when the convoys arrived off the North African coastline at the transport areas. Ships were no longer in the same order they had been in for the voyage across the Atlantic. Plans called for landing craft from one ship to move the cargo of another ship, but in the darkness of the night disoriented navy coxswains reported to the wrong ship or landed on the wrong beach. Some landings went better than others, but all showed that there was considerable room for improvement.[31]

Attack in the West: Safi, Fedala, Mehdia, Port Lyautey

The US Navy and US Army Air Forces supported the Western Task Force landings. At Yellow Beach near Safi, the Western Task Force had to abandon the landing of supplies and equipment after poor beach conditions and a lack of training by navy coxswains caused the loss of seven landing craft. The army and navy had formed beach parties to control landing operations on the assault beaches, but on at least three beaches there was no evidence of functional beach parties. At Fedala, the Task Group commander confirmed that he did not land beach parties with the assault battalions because commanders wanted to use all available boat space for combat troops. Beach parties landed at Safi early in the fight but were slow in executing their duties because they took cover from sniper fire rather than flushing the snipers out and getting back to work on the beaches. In all regards, the beach parties failed to perform at needed levels due to a general lack of leadership, planning, and training.[32]

The lack of experience in amphibious landings played itself out at Fedala's Red Beach, when enemy fire caused the US Army assault force to request a halt in the landings. Landing craft milled about the transport area for hours with no clear direction of what to do or where to go. Even though enemy fire affected just one small area, the navy made no effort to continue landings of troops or supplies on beaches that were not under fire. All of this occurred despite the fact that sea conditions during the November 8 assault off of Casablanca were as flat as a pond—a rare condition for the western coast of Morocco.[33]

The lack of a shore-clearance capability resulted in supplies and equipment landing at wrong locations and many of the supplies loaded on the beach ending up in disarray. The transports had been loaded with a cross-section of supplies so that any one item might not be lost if a transport were lost, but the result on the beaches and at the ports was a mixing of supplies with little idea of what items were where. The shore party had neither the workforce nor the vehicles to remove matériel from the beaches and ports to inshore dumps. The fundamental issue was that no one had realized just how big a task the clearing of the beaches would be and as a result had not planned for the right force to take care of this job. Many of the items that did make it to the beaches were unidentifiable without breaking into the outer containers. Some crates had no markings, while others had inadequate labels, such as "cannon ammunition." Search parties from the combat units scouring the beaches in search of specific items, such as high-explosive ammunition for a 105-mm Howitzer, had a difficult time finding the needed supplies and then quickly transporting those supplies to the front.[34]

A number of ships were out of position during the landings, which caused assault craft to land on the wrong beaches or become stranded on reefs. Inexperienced operators damaged many of the landing boats at Fedala, even those that landed on perfectly good beaches. Other boats found themselves stranded on the beaches and remained there because no tugs accompanied the force. Soon 225 craft of the landing force were stranded on just two of the Western Task Force's beaches. Of all the landing craft destroyed or damaged at Fedala, not a single one was hit by enemy fire. Many boat crews abandoned their boats, thus wrecking vessels that they otherwise could have saved.[35]

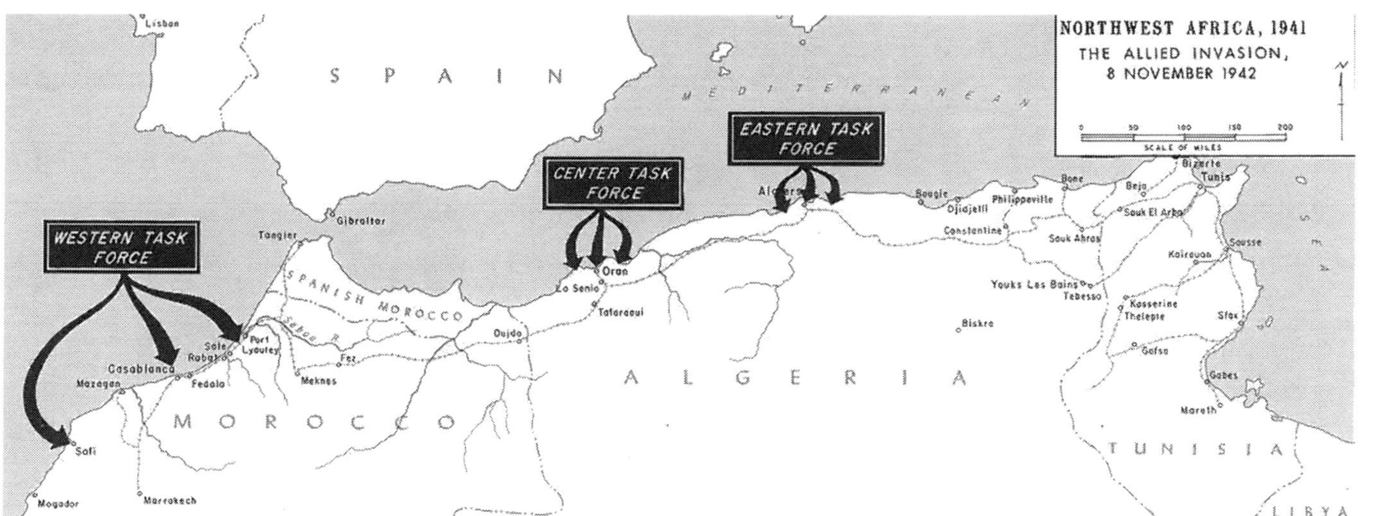

Map 3. Operation Torch, November 8, 1942 (US Military Academy Department of History, West Point, NY)

The French navy did attempt to interdict the landings but had little effect. French destroyers made two sorties out of the port of Fedala on the morning of November 8, but US cruisers, destroyers, and aircraft beat back both attacks. By the evening of D-Day, the Western Task Force had eliminated all French naval opposition.

The French had not expected a landing along this area because of the normally high levels of surf along the Atlantic beaches. The highest-ranking French officer in North Africa, Admiral Jean-François Darlan, did not expect an invasion because he didn't believe the Americans could stage a major assault across the Atlantic. As a result, each of the few French infantrymen that the Western Task Force encountered had an average of fifteen rounds of ammunition—hardly enough to stop an invading army. There was no centrally organized defense by the French along the beach. Despite the challenges in unloading the assault force, Fedala was in US control by the end of D-Day, November 8.[36]

On November 9, D+1, the Western Task Force set out to capture Casablanca. The 3rd Infantry Division (Force Brushwood) started the attack toward Casablanca at 7:00 a.m. with a four-battalion front and initially encountered no opposition. Small French mounted patrols did offer slight resistance as the force approached the city, but this was of little consequence. At 2:00 p.m., however, General Anderson had to delay the attack until additional food, ammunition, and transport were in the hands of the front-line units. Even though there was little opposition, Anderson's units had run out of the food, water, and fuel needed to proceed toward their objective. The Jeeps and small cargo trucks that accompanied the force were proving inadequate for the task; this was the first consequence of the decision to reduce the accompanying truck fleet. Supplies were slowly making it to the shore, but at rates slower than planned, and there was insufficient transportation to move supplies and troops farther inland.[37]

Some French artillery did engage the Western Task Force, but the Allied patrols quickly silenced the offending batteries. The French air force continued to harass the invasion force, strafing the assault columns and bombing the offloading beaches. Collectively, the air attacks, shortage of trucks, slow offloading of supplies, as well as five- to six-foot waves slowed down the rate of the task force's advance until it eventually stalled late in the afternoon.

Anderson ordered the attack resumed at midnight, the morning of November 10. By this time, the force had established an ammunition dump at Fedala and had scrounged enough trucks to resupply the combat units. The US forces encountered enemy patrols on the outskirts of Casablanca but experienced only a few casualties. French field gunfire increased the closer Force Brushwood came to the city. Eisenhower sent a message to Patton: "The only tough nut to crack is in your hands . . . open it quickly and ask for what you want."[38]

The French finally offered a determined defense of Casablanca on the morning of November 10, with 3,600 infantry and ninety guns. Two French corvettes supported the defensive fire, until the US ship *Augusta* and four destroyers pushed the French back into the harbor. By 5:00 p.m., US ground forces were within 400 yards of the city, and the French were ordering their units to pull back. By the evening of November 10, Admiral Darlan issued orders to stop all resistance. Anderson called off the US attack planned for D+3, and the city surrendered.[39]

Conditions near Mehdia were little better. By the night of November 8, the rising surf had stranded about half of the landing craft. In addition, the steep topography of the beach meant that only tracked vehicles could make it inland. The beach at Mehdia had a high escarpment that made the landing of vehicles very difficult. Added to this, there was no exit from the beach, so engineers had to build a road—a task that carried on into the night of D-Day. Had the 60th Infantry Regiment needed its heavy weapons to support an attack on Kasbah that day, those weapons would have been stuck on the beach, unable to reach the advancing infantrymen.[40]

As at Fedala, the units attacking Mehdia landed on the wrong beaches, and supplies were piling up so fast that the work parties had a hard time keeping the piles above the high-water mark. Beaches lacked organization, while stranded sailors wandered about aimlessly. The landing on the north beach occurred five miles north of where it should have been, delaying operations by two and a half hours. One report noted, "Conditions on the beaches during the night presented a scene of indescribable confusion. Surf was rising so fast that about half the craft landing were unable to retract. The delay at the beaches allowed the French to reinforce the stonewalled fort at Kasba, tripling the number of defenders."[41]

High surf on November 9 caused the navy to suspend beach operations. Major General Lucian Truscott, the 3rd Infantry Division commander, made repeated appeals to the navy to resume offloading operations. Landing operations finally resumed at 6:30 a.m. the following morning. On November 11, the Western Task Force opened a secondary landing area at the airport, which provided relief on the beaches but required an additional ten-mile trip for the landing craft.[42]

Despite having the calmest sea conditions of the past sixty-eight years and limited enemy activity, the part of the fleet supporting the Western Task Force damaged or destroyed 242 of the 378 landing craft used in the operation, leaving 162 of these 242 scattered across the beaches. After-action reports about the landing blamed these losses largely on the lack of training and experience of the coxswains piloting the craft, and the navy seemed to agree. The coxswains lacked the training to operate the craft in heavy surf conditions, so they often beached themselves too high on the beaches or hit underwater obstacles. In addition, the landing craft available for Torch lacked sturdy construction, so they could not handle twenty-four-hour operations or contact with underwater rock formations. The lack of recovery craft prevented the navy from dragging beached craft back into the water. High surf and breaking waves breached these stranded craft, where the pounding surf destroyed them.[43]

These losses had both immediate and long-term consequences. The short-term cost was degradation in the ability to quickly land supplies across the beaches and build initial depots to support the assault forces. By November 9, the surf had increased to four to six feet, which slowed down the landings of supplies. With the loss of more than half of the landing craft, ship captains had increasingly to rely on available piers at the seaports, which were of limited quantity and often distant from the fighting forces. This reliance on seaport piers increased the demand for already limited ground transportation to distribute matériel and thus slowed down the resupply of forces.

The lack of an able support organization limited operations at the ports as well. Initially, there were too few trained stevedores to unload the number of ships waiting at anchorage. In addition, there were no corresponding supply and transportation units available to move the offloaded supplies from the piers inland to a supply dump. Ammunition, food, and other items soon clogged the piers because no one was there to take responsibility for them. The *Calvert* took five days to unload—twice as long as it otherwise should have taken. The task force's lack of support organization and control was so total that soon after the clearing of the port at Fedala, a French merchant ship, the *Lumerla*, docked and began to unload its civilian cargo even though two American ships containing cargo trucks and other essential supplies were kept waiting in the harbor. On November 11 and 12, no American ships were brought into either Fedala or Casablanca harbors even though this was a critical period in which the task force commander was trying to establish support capabilities ashore to resupply combat units. It was not until after November 15, one week after the assault, that the first signs of order began to appear at Western Task Force ports.[44]

However, not all of the problems experienced during convoy unloading were the result of poor planning or a lack of service personnel. Equipment problems and poor luck played a role as well. At Safi, the 2nd Armored Division was unloading a medium tank from the *Lakehurst* on the afternoon of D-Day. As the crane hoisted the tank out of the hold, a mechanical problem caused the crane cable to freeze, preventing the cable from going either up or down. The problem took five hours to rectify. At the same time, when the *Titania* raised a light tank out of its hold, the crane cable snapped. The crew spent seven hours trying to find a replacement cable before unloading could resume. These problems occurred as Brigadier General Hugh Gaffey was urgently requesting tanks for the front to help stop approaching French forces.[45]

Through all of this mayhem, George Patton was not amused. On D-Day, the Western Task Force commander had spent eighteen hours on the beaches outside Casablanca trying to motivate the troops to clear the beaches. On D+1, November 9, Patton wrote in his diary, "The beach was a mess and the officers were doing nothing." As of November 30, D+22, there were still piles of gasoline and bombs on the docks, but they gradually dwindled.[46]

On December 23, six weeks after the landings, Patton noted that the docks were at last "in really fine shape."[47] The tragedy is that it took the Western Task Force forty-five days to straighten out a situation that would have been avoidable if a small port unit with trained stevedores and adequate transportation had been included in the initial convoy. This chaotic situation represented a failure of command by task force leaders. From Eisenhower on down, US officers had failed to plan for the landing of a balanced assault force—one that could not only perform its combat mission but also *sustain* itself from the moment it landed on the beaches. This was a critical lesson to learn. Had this same situation occurred facing a determined adversary, the entire invasion force might well have been pushed back into the sea.

The Center Has Its Own Challenges: Oran and Arzew

The Center Task Force was supported by the British navy and British air forces. Despite prior British experience in the war, similar problems occurred in the landings of the Center Task Force. Units were split up, landed on the wrong beaches, and became separated from their equipment. The story of the 48th Surgical Hospital is illustrative of what happened to many support units. Before the hospital had even sailed, six officers, eight nurses, and twenty enlisted personnel were detached from the unit. The remaining hospital's personnel sailed on one ship, while their equipment was in another ship in the convoy. The surgeons and other personnel made it ashore near Arzew during D-Day, but they landed apart, across three miles of coastline. None of the unit's medical equipment made it to the beaches.

Doctors and nurses spent the first night on the beach in foxholes and slit trenches and in helping out at a nearby medical clearing station. The unit eventually collected itself by noon the following day. The equipment was still missing, so hospital personnel went to work collecting and borrowing surgical instruments, equipment, and supplies. Dressings, narcotics, and sterilization equipment were in short supply until D+4, when the unit's equipment finally arrived. Casualty collection, clearing, and dental care were "nonexistent." The lack of capable medical units and the lack of available transportation made the evacuation of the wounded almost impossible, and very few of the wounded received adequate treatment.[48]

By D+6, the most challenging problem facing the Center Task Force was burial of the dead. The II Corps quartermaster had initially requested a graves-registration platoon of one officer and twenty-four enlisted for the invasion force, but the corps operations officer (G3) denied this request on "the grounds that only combat forces were important." Despite the lack of these specialized service forces, an assistant corps quartermaster and some engineer troops eventually established a temporary cemetery for the more than 400 American soldiers awaiting burial.[49]

The lack of trained laborers and an ineffective command organization on the beaches and at the piers proved to be the largest problems of the assault phase of the operation. However, the need for an organizational element to take control of the beaches was not a surprise. The US War Department had identified this requirement before Torch in discussions with the British. A new type of unit was thus formed—the Engineer Amphibian Brigade. It was a multifaceted unit, comprising a boat regiment, a shore regiment, and signal, medical, and supply units. These brigades were supposed to land equipment and supplies as well as receive these items on the beaches and move them to inland dumps. Unfortunately, although the concept proved to be a good one, it was poorly executed.[50]

The assault force lacked both the detailed planning and trained units necessary to carry out an effective landing and unloading of equipment and supplies. Captain James W. Whitfield of the *Calvert* summed up the problems by noting that there were no cooperation, no organization, and not enough personnel and that the officers in charge were both incompetent and too low in rank. Landing with the Center Task Force, the Amphibian Brigade concept proved to be harder to carry out than initially planned. There were no US landing craft available, so the task force told all of the brigade's units to work onshore, even though many were untrained for this duty. The landings operated using British command-and-control systems, and British crews operated most of the landing craft. A lack of training, unfamiliarity with the waters, and darkness produced a scene of chaos and confusion as boats returned to the wrong ships and landed on the wrong beaches.[51]

The brigade was supposed to use its trucks to transport supplies to the dumps, offload them, and then have the supplies sorted, cataloged, and stacked. In reality, the engineers moved supplies to the closest dump regardless of type of item they were carrying and dumped cargo at the water's edge. There was no coordinated control over the landing area and little security for stocks. There was no discipline in the supply area near the beach, and few supplies ended up at the right location. Only gasoline storage went well because the gasoline supply company had its own trucks and hauled gas from the landing areas directly to the fuel dumps.[52]

Despite the chaos, ship captains did what they could to offload essential supplies. One such captain resorted to towing life rafts behind the landing boats in order to get supplies of gasoline ashore—each life raft holding 136 cans of fuel. Shore parties of engineers succeeded in establishing dumps and aid stations along many of the beaches. Ammunition, rations, water, and fuel began to accumulate in small quantities; however, movement of supplies forward from the beach dumps to the fighting units was almost nonexistent. Supplies such as water, fuel, and ammunition weighed a great deal, distances were too far to carry supplies by hand, and there was a lack of ground transportation. Supplies made it ashore before the trucks. Planners had counted on Jeeps and half-ton trucks to haul supplies, but these small vehicles proved inadequate for the task. Supplies continued to be unloaded at all beaches with no real direction regarding their placement.[53] By the end of the first day, only 39 percent of the troops, 16 percent of the vehicles, and just a little more than one percent of the supplies had made it ashore.[54]

In general, the overall lack of service units in the task forces precipitated the problems on the beaches and at the ports. For example, on the first two convoys the Center Task Force included only 260 supply personnel in a total force of more than 60,000. The II Corps quartermaster asked to include additional supply units, but the task force commander denied his request on the assumption that local civilian labor would be available in North Africa to offset the needed units. Upon landing, however, the task force discovered the local labor to be limited, unskilled, and in need of constant supervision. Poor assumptions and a lack of understanding concerning the challenges of supporting a deployed force led to widespread problems that were more than anecdotal.

Meanwhile, in the East: Algiers

As in the center, the Eastern Task Force was supported by the British navy and air forces. However, in the Eastern Task Force sector the combined US and British landings experienced fewer problems, in part due to experience, training, and rehearsals in ship-to-shore landings conducted in Great Britain. The lack of a large organized ground resistance from the French helped as well. Beaches in the task force sector spanned roughly 50 miles, from Castiglione in the west to Surcouf in the east.

Landings in the western part of the task force area went generally smoothly, despite rough surf, and the British 11th Infantry Brigade encountered no resistance. To the east, near Fort Duperre, the British 6th Commando had a few challenges. Breakdowns in landing craft and other delays put the final landings more than five hours behind schedule. Landing craft carrying the US 168th

Combat Team became misoriented and were carried off course, scattering the unit. Across all the beaches, commanders and men were forced to improvise as plans failed to match reality.[55]

To the east, the American 39th Combat Team of the 9th Infantry Division found similar challenges as the other forces in getting the assault forces organized for the initial landings. However, these challenges were quickly overcome, and by early morning the infantry battalions were moving toward their objectives and encountering limited resistance.

As with the beaches of the Western and Center Task Forces' sectors, the Eastern Task Force quickly found its beaches chaotic. Supplies were dumped with little thought or plan. Keys were missing for vehicles. Fortunately, an anti-Vichy coup in Algiers resulted in little organized resistance to the invasion, and the city was largely in Allied hands by the evening of November 8, with the US 39th and 168th Regimental Combat Teams and the British 11th Brigade Group ashore. As operations continued, however, British beaches had only a limited throughput capability for supplies and equipment, which slowed the buildup of the British First Army.[56]

All supplies for the Eastern Task Force came from Great Britain. Common items, such as food and water, were of British origin. Items issued solely to US forces, such as repair parts for American equipment, came from stocks maintained by the SOS in England and delivered by the British navy and army.

A major task for the Eastern Task Force was to begin establishing a number of airfields for use by the Eastern Air Command once fighting subsided. Within seven weeks, the Allies hoped to station twenty-five squadrons totaling 454 aircraft in the remote areas surrounding Algiers. Whether they could accomplish this depended on how quickly the logistics support structure could offload ships and move matériel inland. Items such as steel matting, fuel, and bombs took priority as the theater began to transition from securing the beaches to preparing for major combat operations against the German and Italian forces to the east.[57]

Consequences

The Axis response was quick and deliberate. Hitler made the Mediterranean a priority upon hearing of the landings and ordered immediate reinforcement of Tunisia. The Luftwaffe stepped up operations, while the German navy focused efforts on sinking Allied ships. Axis ground reinforcements began arriving in northern Africa in early November.[58]

The loss and damage to so many Allied landing craft resulted in a reduction in the overall numbers of assault craft available to support national strategies. Numbers of assault craft in 1942 and throughout the war never met the global demand for this unique capability. The United States was balancing the allocation of both funds and materials, such as steel, among many competing programs. The loss of any landing craft meant that the Allied global capability to support amphibious assaults was in some way hindered. Indeed, the loss of landing craft affected not just the Mediterranean and European theaters but the Pacific theater as well. Amphibious operations were one of the essential characteristics of World War II, and all the theaters were demanding increasing numbers of assault craft. The issue was so serious that the loss of landing craft became a topic of discussion at the Anfa (Casablanca) strategy conference in January 1943 (named for the hotel in which the conference was held), and Roosevelt and Prime Minister Winston Churchill would wrestle with the issue throughout the remainder of the war.[59]

Ultimately, the surprise achieved by the landings prevented any coordinated response by the French military in Morocco or Algeria. French resistance was spirited in select areas but had little real impact on the landings. The Allies got lucky because a coordinated French sea and air attack on the beaches might well have prevented the Torch forces from achieving their desired objectives. On November 10, Admiral Darlan, the senior French officer in North Africa, broadcast the order for all French forces in North Africa to cease hostilities. By 7:00 a.m. the following day, the front was quiet, and the Allies then shifted their focus to the capture of Tunisia and began working out the details of how to integrate the French military into the Allied coalition. As of November 12, it appeared that the Axis nations were not willing to occupy Spain, and there was little threat from Spanish Morocco. Eisenhower was now free to reorient the Torch forces to the east and to begin a race with the Germans to see who could seize and hold Tunis first.

The Allies had much to learn before attempting the next amphibious landing. Aside from the amateurish landings made thus far, the most significant issue was a fundamental misunderstanding of how air forces could effectively provide close air support to ground forces. Fortunately, the weather and the enemy generally cooperated—otherwise, the landings could have been much worse. As it was, the three task forces made it ashore and began building up combat power in preparation for the next phase of the operation. Now they needed to build a viable sustainment capacity that could quickly support the drive into Tunisia, the true goal of the operation.[60]

Eisenhower noted that the "chief hope of anticipating the Axis in Tunisia lay in our acting with utmost speed."[61] The task forces were ashore, but before any large-scale offensive operation could begin, they needed to build up a support capability that had the ability to sustain long-term operations over hundreds of miles in a harsh and undeveloped environment. The initial combat force secured the initial objectives near the beaches, but it required reconstitution before the divisions could move beyond the initial lodgment areas. To accomplish this required a capable sustainment force with the right equipment and supplies.

Even though the Western Task Force had an assigned logistics element, SOS Task Force A, Patton and his planners did not integrate the SOS into the D-Day landings. The Atlantic Base Section commander, Major General Arthur Wilson, noted later, "There was no consideration given to the problem of supply in the Western Task Force beyond the landing of troops on the beaches." The problem was not just with the SOS; the D+5 convoy, including the task force G4, Colonel Walter Muller, didn't arrive in Morocco until D+11. Because of this seemingly intentional effort to relegate the task force logisticians and support elements to later convoys, the Western Task Force's ports were chaotic, confused, and unsupervised. Port facilities in the area of Casablanca were available soon after the landings; however, the lack of trained port personnel prevented the task force from taking full advantage of these facilities. As a result, ship offloading at the ports was slow and largely unsupervised. Stacks of supplies clogged docks because there were no SOS forces or equipment available to make sense of the items that had been unloaded and to move these supplies from the ports and beaches into depots and dumps farther inshore. The ports and beaches had become chokepoints.[62]

By the evening of November 16, there was a continuous pile of miscellaneous supplies on the piers of Casablanca that averaged 10 feet high, 20 feet at the base, and approximately 700 yards in length. One section included a lethal mix of steel mat, ten-gallon containers of aviation fuel, Vienna sausages, .45-caliber ammunition, lubricating oil, and a cloverleaf of 105-mm incendiary ammunition. Few people were working to clear the mess, and the follow-on convoy would only be adding to the chaos.[63]

The first convoy to follow the D-Day invasion force was the D+5 convoy. This collection of nine troop ships and eleven fast cargo ships arrived in Casablanca on November 18, 1942, only to find confusion and disorder at the city's docks. Members of the 6th Port, a unit on the D+5 convoy, which specialized in port operations, found conditions at the Phosphate pier looking "as though some gigantic overhead scoop full of supplies had suddenly emptied its contents. Apparently, nothing had been hauled away and nothing had been stacked. One box was simply on top of another. On the other dock we could see boxes, crates, ammunition, and gasoline drums piled up and scattered from one end to another."[64]

Port operations were new to the US Army; units such as 6th Port were less than five months old. Commanded by Colonel Hunter Clarkson, the 6th Port landed on November 19 and immediately went to work. The unit was new, but the men in it had a wealth of experience as civilian stevedores, dockworkers, rail operators, and truck drivers. When supplies and equipment came ashore, they were piled up on the docks, even while the combat forces were moving inland and calling for more fuel and supplies. There were not enough trucks, rail cars, or service units to clear the ports. Adding to the difficulty of the task was the tremendous amount of rain that fell, turning hillsides into cascades and valley floors into lakes. Everyone in the command, including cooks and clerks, was put to work to clear the docks. A typical shift was sixteen to eighteen hours long, and meals consisted of cold C-rations. The rains had knocked down most of the pup tents, so many men chose to sleep in the holds of ships—anywhere that was dry. The Casablanca port scene did include trained stevedores who knew what they needed to do, but they had to contend with a larger group of untrained beginners and local Arab labor. The battle with the French navy had wrecked all of the port's cranes. However, by February 1 conditions had improved, and the Allies were employing 8,000 local Frenchmen and Arabs to work as stevedores.[65]

Units such as the 6th Port began work to make sense out of the docks and establish a storage capability away from the piers; however, the shortage of cargo trucks hampered these operations. Few military trucks were available, so US forces used any readily available local means of conveyance, such as horse-drawn wagons and charcoal-burning trucks. One of the ships, the *Lorraine*, carried forty trucks and was the third ship offloaded at Fedala, but due to waterproofing issues none of the vehicles was serviceable.[66]

The shortage of trucks was a hindrance felt across the force. Planners had allocated 200 trucks for the unloading of supplies at the ports, but less than half of them were in French Morocco when the D+5 convoy arrived. Additional trucks arrived with the D+5 convoy but could not be unloaded until workers cleared the ports. Patton's combat units in the Western Task Force had priority use of the trucks that did make it to theater, meaning that the service units had to make do with whatever was available.

Besides deciding to keep many of the cargo trucks in New York due to limited shipping space, Patton decided to substitute smaller quarter-ton trucks (Jeeps) for the two-and-a-half-ton cargo trucks that were authorized and available for the force. On the one hand, this decision made sense given the shipping restrictions that the Western Task Force was wrestling with; on the other hand, the decision to deploy with smaller trucks had operational and logistical implications. Without the two-and-a-half-ton trucks, affectionately known as "deuce and a halves," Patton's force lacked the ground transportation necessary to resupply or transport itself. The problem was simply a matter of physics. In terms of cargo capacity, one deuce and a half equaled ten Jeeps. Patton had focused on weight and space, not on capability or need, when he had made his decision about equipment priorities.[67]

The combined air force element of Torch, the Moroccan Composite Wing, found itself stranded on the beaches following the landings. As with other units in the Western Task Force, Patton had decided to leave most of the wing's automotive equipment back on the piers at the US ports of embarkation, expecting the wing to rely on the ground forces for the movement of its personnel and supplies. However, because the ground force was *also* short of transportation, there was little movement of bombs, ammunition, and fuel for the air corps once it reached Morocco.[68]

The Composite Wing found itself in a precarious position largely because wing planners consciously focused their efforts on ammunition requirements and decided to leave all planning for spare parts and automotive equipment to the Western Task Force. As such, there was no liaison between the wing and the task force on the wing's specific transportation needs once it arrived in North Africa. Patton treated the wing's truck assets the same as any other unit's assets, apparently not considering the impact this would have on his air support. Either no one argued for the reinstatement of the vehicles, or Patton failed to grasp that in some situations trucks could be more important than infantry and armor.

The decision to eliminate a large number of cargo trucks from the initial assault convoys was not just a US military oversight. British planners likewise decided to keep a large number of trucks back in Great Britain due to the shortage of motor transport vessels. This lack of trucks in all three task forces meant that it would take longer to clear supplies off of beaches or ports and longer to establish a significant distribution capacity and that large infantry formations were limited to movement by foot or rail. One story illustrates just how much of an impact this decision had on the force. The deputy director of supplies and transport for the British First Army estimated that he could support forces only up to 100 miles east of Bône until additional trucks arrived on the D+32 convoy. This estimate not only proved to be accurate but also played a key role in the British failure to secure Tunis before the German military had an opportunity to reinforce the region.

By November 16, the British First Army was 50 miles from Tunis. The advance stalled, however, due to the combination of resource constraints, weather, and increasing Axis strength. Both the British and US logistic systems were stretched to the limit. The combination of limited port throughput, limited transportation equipment, a single rail line, and washed-out roads slowed all Allied forces regardless of national origin. An Allied attack on November 24 failed, and German counteroffensives on November 27 and December 1 forced an Allied withdrawal. Eisenhower decided to wait out the winter and provide time to build his force.

The limiting factor in the fight for North Africa was shipping capacity. The US Navy could provide only so many escort vessels, which limited the number of cargo ships in the convoys across the Atlantic. In prioritizing cargo for the reduced number of ships, commanders cut down the number of cargo trucks. Had additional convoy escort vessels or trucks been available, the campaign in Tunisia could possibly have ended five to six months earlier. However, if French forces in North Africa had put up a stronger fight, the real need would have been combat forces. Nonetheless, Eisenhower and his planners knew that the landings would start a race with Germany to see which side could land reinforcements most quickly. Eisenhower wrote afterward, "Our chief hope of anticipating the Axis in Tunisia lay in our acting with utmost speed. Indeed, we were involved in a race, not only against the build-up of enemy forces, but against the weather as well."[69]

The problem for the Allies in late 1942 was not so much a lack of combat force in North Africa but the lack of a means to move that force quickly around the theater. Limited rail lines, poor roads, and the lack of ground transport vehicles contributed to a situation in which AFHQ could not respond to opportunities. In November 1942, the Germans lacked the combat and logistical strength to repel a strong Allied push into Tunis; however, the British also lacked the means to rapidly deploy and sustain a large force across the 220 miles from Bône to Tunis. Likewise, Patton's Western Task Force was largely unengaged in French Morocco, but the Americans lacked the means to move divisional-size formations to the east. The Allies had an opportunity before them, but they lacked the means to take advantage of it. Had sufficient transport been available in December, the Allies might well have seized Tunis that month, and the Germans would have had their supply chain interdicted—all of which might well have prevented the battle at Kasserine Pass in February 1943. This account is hypothetical, but it does serve to illustrate how decisions on issues as mundane as trucks contributed to the outcome of the war's battles and campaigns.

Eisenhower believed that the drive for Tunisia failed for three reasons. First, limitations in shipping prevented the force from having sufficient strength to deal with the distance and force it faced. Second, shortages of trucks and the limitation of a single-line railway slowed down the rate of advance. Third, unseasonable rains stalled air operations. Together, these three factors created a situation that forced the Allies to pause for the winter, but the pause did allow an opportunity for the service units to build up the theater into an effective base of operations.[70]

2

Establishing the Theater

Keep all wheels rolling. . . . It is the results that count.
—Brigadier General Arthur Pence, Eastern Base Section Commander

Following the landings in North Africa, each task force handled support responsibilities in its own respective area of operations. Allied Force Headquarters provided overall guidance, but there was no formal communications zone or single support agency. Each base section took care of its own task force and anyone passing through the area.

Although weak at first, the base sections steadily increased in strength with the arrival of additional convoys. All three task forces lacked organization. Ports and supply dumps were in disarray. Pilferage by the population had become a major problem, leading to the requirement that a US soldier accompany every wagon or vehicle. Infantry units formed bucket brigades in order to move supplies off the docks. Vacant fields and broken-down warehouses served as impromptu supply dumps and storage locations for the thousands of barracks bags separated from their owners on the invasion force. Military clerks and mechanics served as guides in an effort to direct the endless stream of vehicles from the docks to the supply dumps, but even with that effort, many drivers became lost.[1]

To complicate the supply situation further, the Western Task Force had planned to allow partially unloaded vessels to return to the United States if there was insufficient time to empty them before the convoy had to make room for the next convoy. The convoy schedule was extremely tight, and any delay would affect numerous other convoys. Unfortunately, nobody had thought of inventorying these partially filled ships prior to their return from the Mediterranean. As a result, no one was entirely sure of the exact quantity of matériel that had made its way onto the shores of North Africa. This issue was a problem that the base sections would have to resolve as the theater began to consolidate and AFHQ formed a communications zone.[2]

Innovation and adaptability proved to be essential elements in the theater's ability to support its forces and operations. Engineers removed the ends from oil drums to serve as makeshift forms for concrete footings. Medics used red iron ore to create a red cross against a bed of limestone at hospitals to protect them from enemy attack. The shortage of lumber resulted in walls for buildings made out of paper, tar, and chicken wire. Units throughout the rear area combined imagination with available resources to compensate for any shortages in supplies and equipment. This innovation became an integral part of the force and helped the US forces overcome adversity not just in North Africa but also throughout the Mediterranean theater.[3]

Resources such as merchant shipping and convoy escorts were scarce throughout the world, so the War Department and AFHQ needed to find efficiency wherever possible. Issues with limited shipping space reached the highest levels, so that even before the convoys had sailed for North Africa, the Combined Chiefs of Staff had directed that "it is highly desirable to save shipping space and many steps have to be taken to accomplish this." The twin-unit pack (TUP) was one such innovation that addressed the worldwide shipping shortage. It was simply two complete vehicles, disassembled and packed together with all the necessary parts for assembly.[4]

The US Army's Ordnance Department had originally planned to ship all trucks to the theater in a fully assembled configuration; however, the army and General Motors developed a plan to ship these same trucks disassembled and then assemble them later after arrival in the Mediterranean theater. Due to their large size, assembled trucks tended to fill up a ship before the ship's weight limit was reached. As such, ships filled with assembled trucks sailed at less than full weight capacity. By loading trucks in a TUP configuration, the loadmasters at the New York Port of Embarkation could better utilize the full cargo capacity of each ship up to the overall weight limit. In short, the army could load more vehicles in the same amount of space, which resulted in a greater transportation capacity within the theater. The only problem was that the trucks required reassembly, and the deserts of North Africa lacked both the manufacturing capabilities and the trained workforce of industrialized nations.

The solution to the assembly problem was the use of specialized units, such as the 302nd Ordnance Regiment. This regiment contained skilled volunteers who had attended specialized training at large US industries prior to sailing for the theater. The army and the Munitions Board granted a triple-A priority rating for the project, and the first convoys containing TUP vehicles arrived in Casablanca and Oran by the end of 1942.[5]

Each TUP consisted of four boxes: one contained two truck frames with the motors attached; the second box contained the cabs and wheels; the third held axle assemblies; and the fourth contained the truck bodies. Overall, shipping one deuce-and-a-half TUP, compared to shipping two fully assembled vehicles, saved 59 percent cargo space. Manufacturers provided assembly plant layouts and any specialized equipment necessary for vehicle assembly. Planners anticipated an assembly rate of 100 vehicles per day for each assembly line.[6]

Figure 1. General Giraud visits the TUP-SUP site (Military History Institute, Carlisle, PA)

The assembly line at Oran was located where 120 TUPs had been dropped in two muddy fields because there was no heavy-lift equipment available to move the heavy boxes, each of which weighed up to 10,000 pounds. To assemble the vehicles, ordnance units used open fields and four-ton wreckers in lieu of the concrete floors and heavy cranes envisioned by the manufacturers. The demand for the trucks was so great in the Center Task Force area that many units received vehicles lacking even a road test.[7]

The complete assembly line gradually improved over seven weeks. Engineers erected buildings with overhead hoists, drainage, lighting, water, and compressed air. In one single day, the line at Oran assembled and delivered 143 of the two-and-a-half-ton trucks. Installation included mounting of cabs, lubrication, installation of wheels and wiring, and fabrication of vehicle bodies. This was an impressive effort, which, although it occurred largely in the background, provided the theater with a critical item of equipment. Eventually, each assembly line occupied more than a million square feet of space.[8]

One of the greatest contributions to the theater was the improvement of the North African rail network. When the Allis landed in November 1942, the rail system in North Africa presented a daunting list of challenges. Tracks were typically limited to a single line, thus obstructing two-way movement. There was no one standard of track—the Chemin de Fer de Maroc, the Railway of Morocco, was standard gauge in the West, but the line changed to narrow gauge in eastern Algeria, east of Ouled Rahmoun and Tébessa. In addition, there were few locomotives and rail cars; those that did exist typically had relatively little capacity compared to the demands the military needed to place on them. The rail system needed a great deal of work to maximize its potential.[9]

Accompanying the invasion force on D-Day was an advance element of the 703rd Railway Grand Division. Formed along a traditional railroad organization, the Railway Grand Division replicated a civilian general superintendent's office and operated three to four railway battalions, a shop battalion for heavy maintenance, and a base depot company for supply. Each rail battalion included a headquarters company, which dealt with dispatching, supply, and signals. Company A handled maintenance of rail lines and facilities along the lines, while Company B operated the roundhouse and repaired rolling stock and locomotives. Company C consisted of fifty train crews and a train master. Commander of the Military Railway Service was Brigadier General Carl R. Gray Jr., the vice president of the Chicago, St. Paul, Minneapolis & Omaha Railroad before the war.[10]

By November 18, 1942, the first personnel of the Railway Grand Division arrived and began making contact with the French railroad operators to arrange rail movement of all supplies and equipment arriving on the D+5 convoy. Since the rail lines technically belonged to the French, the Allies had to negotiate for use of the lines and rail equipment. As luck would have it, George Falson, a US veteran of World War I, had been living in Casablanca before the invasion, so he grabbed his discharge papers and met the assault units on the beaches of Fedala when the assault boats came ashore. Falson was very familiar with how the French rail system operated in North Africa, so he attached himself to the US Railway Division as "interpreter extraordinaire." This local assistance helped US forces acquire locomotives and other rail support from the French and eventually helped the US forces assume all switching service.[11]

The War Department shipped rail locomotives from the states in a partially dismantled configuration. They arrived in Morocco, where soldiers then transported the disassembled locomotives 630 miles east to Sidi Mabrouk, Algeria, for reassembly by the 753rd Shop Battalion. The Shop Battalion finished reassembly of the first two locomotives in seventy-six hours and then loaded the railroad engines on a train for movement to Ouled Rahmoun, where the locomotives went to work on the narrow-gauge tracks.[12]

German forces quickly realized the importance of the single rail line running along the coast and worked to interrupt rail service. German special-operation elements landed by parachute behind the front lines to sabotage the rail lines, while German pilots targeted rail facilities and supporting infrastructure. The rail battalions had an ongoing challenge to keep bridges, rails and switches, and damaged machinery repaired. Despite the challenges, the rail units kept pace with advancing combat elements and provided the supplies and equipment needed to win the battle of Tunisia and North Africa.

The long lines of communication represented a twofold challenge for the Allies—how best to operate such resources as the rail system, which was a civilian entity, and at the same time provide security to prevent attacks and pilferage. Pilferage is a problem that exists in most operations, especially situations where local populations have little available resources. An arrangement developed where French civilian agencies would help run the rail system and other civilian infrastructure, wherein the French would deal directly with the Arabs to arrange security. This procedure released combat forces for the front, provided jobs for the local population, leveraged the expertise of French administrators, and proved to be extremely successful.[13]

Formalizing the Communications Zone (Rear Area)

In the period from November 1942 to March 1943, North Africa operations changed into a coherent theater of war. Initially, each task force was responsible for coordinating its own support, but by spring the time was right to transfer to a more effective and efficient system of centralized oversight. In addition, the Casablanca Conference in January 1943 confirmed that operations in North Africa were the beginning of a larger series of campaigns within the Mediterranean as the Allies agreed to the follow-on assault of Sicily. With the decision to invade Sicily, the Allies needed a formal Mediterranean theater with an accompanying headquarters and support structure.

Many resources were restricted due to competing worldwide demands from the different theaters or from shipping limitations. The Mediterranean theater needed a means to make the most of its available resources. Command of combat units shifted as well in order to provide more centralized control over the ground, air, and sea forces. The establishment of the theater represented an effort to bring order to the combat and communications zones, a task necessary to support a long series of operations and campaigns. Without a functioning theater, the Allied Forces would not be able to support their forces deployed overseas.[14]

The divisions had secured their objectives, and now came the task of turning North Africa into a base of operations from which to launch the assault on Tunisia as well other subsequent campaigns. To accomplish this task, the theater needed an organization to coordinate the support efforts in the rear areas, and

Figure 2. General Sir Humfrey Gale (US Army Center of
Military History, Washington, DC)

the base sections needed to establish the installations, warehouses, and repair
facilities to support the force as well as prisoner-of-war (POW) camps. Cooperation among the senior officers and innovation demonstrated by individual
soldiers played key roles as the US military began to build a theater in a very
austere environment. Unfortunately, the lack of clear guidance from Eisenhower
regarding the division of responsibilities for the rear area and command of the
SOS led to considerable confusion and tension.

The failure to capture Tunis in December 1942 provided a setback for Allied
strategy but also an opportunity for logisticians to build up the rear areas in
preparation for a spring offensive. Eisenhower had paused operations, but the
winter of 1942–1943 was a time of great energy devoted to the development of
the task force support capabilities. The armies received new equipment, and
supply dumps built up reserves. Support forces took advantage of the lull
to refine the systems needed to effectively move and support the force. Major

General Humfrey Gale, the British and AFHQ chief administrative officer, arrived from London on December 1.

Gale had a vast amount of logistics experience, having served as a deputy assistant quartermaster general and assistant director of shipping and transport in the UK War Office during the interwar years. He had been the deputy assistant quartermaster general for III Corps in 1939–1940, for which he was awarded Commander of the Most Excellent Order of the British Empire. Now, his first task as chief administrative officer was to begin sorting out the mess at the North African front.[15]

The Base Sections

In order to relieve the task forces of defense requirements and supply duties in the rear areas, AFHQ activated two specialized commands, similar to the concept of support exercised during World War I. First, the Mediterranean Base Section (BS) arose on November 10, 1942, with its headquarters at Oran, Algeria. Seven weeks later the Atlantic BS, with its headquarters located at Casablanca, Morocco, was activated on December 30. By the beginning of 1943, the two base sections were no longer under task force control but, rather, directly under AFHQ control.[16]

The Atlantic BS handled support for forces in the area occupied by the I Armored Corps (the former Western Task Force). To the east, the Mediterranean BS assumed responsibility for support of all forces in II Corps (the Center Task Force) as well as support of US forces operating in the area of the British First Army (the Eastern Task Force). The one exception was that British forces were responsible for the support of all US forces operating as part of the British Eastern Assault Force in Tunisia.[17]

Besides having the responsibility for administrative support of combat forces, AFHQ also assigned the two base sections responsibility for all civil affairs matters. In addition, requests from US forces for supplies and equipment went directly to the base sections. The base sections found themselves with a growing list of customers—they now had to support military forces, civilians, French forces, and POWs.[18]

Eisenhower knew that the force was not an occupying one but instead needed to be seen as an accepted ally or visitor. Accordingly, "far from governing a conquered country, we were attempting only to force a gradual widening of the base of government, with the final objective of turning all internal affairs over to popular control."[19] In this situation, the Allies had to work with the local government and populations to establish goodwill and an atmosphere of cooperation. This meant that Eisenhower had to consider the welfare of the people to a greater level than otherwise might have been the case. Support from the people and from their local government was essential if the Allies were to avoid a situation in which the rear areas might be threatened, which would require the stationing of significant amounts of combat forces to provide security. Thus, civil affairs quickly became a critical element of the theater support strategy.

The centers of activity for the base sections were the ports at Casablanca, Oran, and Algiers. Each port had to have sufficient berthing spaces to handle the number of ships arriving with the convoys as well as enough lifting equipment and transportation networks to clear matériel off the piers. Ports and depots needed to be close enough to the front to support Allied strategies but far enough removed from southern France and Tunisia to avoid large Axis air raids.

The two base sections had other wide-sweeping responsibilities that included oversight of all depots, fund estimates for the task forces, control of transportation not within the task force, all construction, signal facilities, and civilian labor. In addition, the base sections had to work closely with the US Navy in all matters involving the clearing of ports and harbors, the operation of all ports and moles, the loading and unloading of ships, the control of all lines of communication, the fueling of ships, and the security of all ports and military installations.[20]

As Operation Torch moved into 1943, the base sections worked to make some sense out of the piles of supplies that had accumulated in the task force areas. For example, the Atlantic BS was faced with a pile of ammunition at Fedala that measured 150 feet deep for a mile and a half on either side of the railroad. Ammunition was "strewn, scattered, jumbled, mixed up—it seemed every type, size, lot number, and caliber ever produced was piled there in the most amazingly involved condition that any Ordnance man or officer had ever seen." This is just one example of the state of the intermediate supply depots in early 1943; all depots needed to be cleaned, organized, and inventoried to take full advantage of the matériel that had been received within the Mediterranean theater.[21]

Without this organizational effort, the base sections would not know what supplies were on hand, where the needed supplies were, or how to issue supplies in an effective and efficient manner. Dumps were often unnumbered. Units had to scour the piers and surrounding areas for supplies believed to be on hand. This situation, left unresolved, could have serious consequences because if a unit needed critical supplies and the base section had no visibility of what was actually available, the base section then had to send the requisition back to the United States for fulfillment. The requisition might require months to fill and would take up valuable shipping space even though the needed items might already be located within the theater and available for use. Finally, the net explosive capability of all that ammunition piled up in one place represented a major safety problem, especially because the chance of air raids was ever present. Fortunately, it does not appear that the German Luftwaffe ever took advantage of these lucrative targets, but the risk was there nonetheless.[22]

Situations like this were common. Support units found supplies and equipment haphazardly strewn about impromptu supply depots, in vacant lots, and alongside a rail line. These piles were an unintentional and normal result of the growth of the theater; however, better planning could have limited the extent of the disorganization. At the time, units on the assault wave simply worked to unload ships as quickly as possible. Workers established the initial supply dumps as a means to expeditiously clear matériel from the ports, but with little organization or record keeping. Plans also changed—the ammunition dump at Fedala was supposed to receive 35 to 50 percent of all incoming munitions for the Western Task Force, but it instead received nearly 100 percent. Conditions deteriorated when subsequent convoys arrived and added their supplies to those that were already on the ground. The shortage of trucks prevented units from quickly moving supplies off ports and rail spurs. The base sections found themselves with a problem that grew with every additional convoy.[23]

Eisenhower begged for additional transportation. The War Department and US industrial base responded, sailing a convoy less than three weeks later loaded with war matériel, including 5,000 trucks, 2,000 trailers, and 400 dump trucks. Between late February and late March, the War Department shipped an additional 24,000 vehicles and more than a million tons of supplies. This was a war of supply.[24]

Administrative Oversight

In January 1943, the situation in the rear was becoming a major distraction for Eisenhower. Losses from combat and accidents had taken a toll, particularly for the air forces. Units were operating below nominal operating strength; the logistics network was at its breaking point. The army chief of staff, General George C. Marshall, arrived in Algiers on January 24 and was not impressed with what he saw. There was a lack of discipline, organization, leadership, and training. Following a tour of SOS units, Marshall insisted that several senior officers be relieved and that there would be no more "goddam drugstore cowboys standing around." Marshall ordered Eisenhower to straighten things out and to shift oversight of administrative responsibilities to someone else.[25]

Although Eisenhower had the responsibility to provide administrative support to all US forces as the theater commander, he did not have the time or inclination to do this job. As the Allied commander in chief, he had more pressing duties, such as to keep the English–French–American coalition together and to deal with

Figure 3. Major General Everett Hughes (US Army Center of Military History, Washington, DC)

various political officials. Eisenhower needed another general officer who could focus on the administration of the theater, so on February 6 he chose Brigadier General Everett S. Hughes to be the deputy theater commander. Army doctrine did not include such a position; however, this same doctrine did not envision the myriad competing demands required of an allied commander in chief.

Everett Hughes was a long-serving officer, initially branched as a field artillery officer after graduating from West Point in 1908. Three years later, he transferred to the Ordnance Corps and after that deployed to Mexico in 1916 as part of the Punitive Expedition. Hughes served in the Allied Expeditionary Force during World War I and taught at the General Staff School at Fort Leavenworth, Kansas, from 1923 to 1928. This mixture of field and teaching experience within the army's professional military education system was common for many of the new war's senior officers.

Hughes was a close confidant of George Patton—the two of them had been a year apart at West Point, and both had served in Mexico and France. Hughes and Patton's diaries reveal that the two generals often got together whenever Patton was visiting AFHQ, with Patton frequently using the opportunity to remark over several drinks on how he thought Eisenhower or other senior leaders, such as the British general Bernard Montgomery, were doing. This was an intriguing relationship between a senior combat commander and a support officer, especially considering Patton's attitudes toward his own task force support commander, Brigadier General Arthur Wilson. One benefit of such a relationship was that Patton had an inside ear and confidant within AFHQ, while Hughes could learn firsthand of any administrative concerns coming from the front.[26]

Hughes was a large man, standing at six foot five inches. In addition to his friendship with Patton, Hughes was also on close terms with Eisenhower, the two having served together in the 1930s. Hughes described the duties of the deputy theater commander simply as being "charged with the responsibility of relieving the Theater Commander of all possible [administrative] details." In effect, Hughes became responsible for the support of all US forces, while the chief administrative officer, General Gale, had the responsibility for all British forces.[27]

Gale and Hughes were the two officers who oversaw theater logistics for the Allies. Hughes handled the American side, while Gale took care of the British, in addition to serving as the senior logistics staff officer in AFHQ. SOS headquarters for the North African Theater of Operations, United States Army (NATOUSA) reported to Hughes; the British line of communication reported to Gale. Constant communications and cooperation between these two men were essential to overcoming the many challenges that routinely arouse.

Because US doctrine did not include the provision for a deputy theater commander, Hughes asked that he also become the commanding general of the communications zone—a familiar doctrinal concept. This additional designation did not come with any additional responsibilities or staff, but it did provide an authority needed to deal with the War Department SOS and helped to define Hughes's duties in terms that were more traditional. Eisenhower agreed and on February 9, 1943, signed a memorandum directing Hughes to "establish, operate, and command a U.S. Communications Zone for NATO [North African Theater of Operations] USA. You will assume all possible U.S. administrative and supply duties now being performed by AFHQ, in order to relieve AFHQ to the maximum of supply and administrative matters applying to U.S. forces. You will also be responsible for the detailed development of supply plans for American forces in future operations to conform to the broad plans of AFHQ. . . . In addition to the duties indicated above you are designated as Deputy U.S. Army Theater Commander."[28]

This assignment became official three days later when AFHQ published a general order on February 12 directing the establishment of the NATOUSA communications zone. With this order, the Fifth Army became responsible for the general defense of the communications zone, and the duty of the communications zone commanding general/deputy theater commander was to coordinate and synchronize the activities of the AFHQ staff, the Fifth Army, and the base sections. Thus, just one man—Everett Hughes—represented the entire rear area.[29]

Unfortunately, Eisenhower's order did not specify the delineation of responsibilities among the different logistics staffs and organizations within the theater. With it, he had achieved his goal of shifting the administrative burden off of AFHQ, but there arose a difference of opinion between AFHQ, the US theater, and the base sections on the roles that each was to perform and who was to report to whom. This controversy continued for at least six months and served as a point of tension among the organizations that depended on efficiency, not disagreement, to succeed. Major General Walter Bedell Smith (Eisenhower's chief of staff), Brigadier General Tom Larkin, and Brigadier General Everett Hughes had different ideas of what the chain of command ought to be for the rear area and the SOS.

One area of contention was the command chain for the rail forces. While the Railway Grand Division supported SOS NATOUSA in carrying out sustainment operations for the North African theater, it did not report directly to the SOS. Instead, the commanding general of the Railway Division, Brigadier General Gray, was "responsible to the Deputy Theater Commander for the wellbeing [*sic*], discipline, and training of the United States' rail-way troops." Larkin could ask Gray for support, but Gray ultimately worked for Hughes. Although this command arrangement finally worked, it was not as clean and efficient as it might have otherwise been. However, considering the personalities of the generals, it probably was the optimal arrangement.[30]

February also saw the creation of a third US base section. The British were still supporting the US forces attached to the Eastern Task Force out of the Port of Algiers. This responsibility increasingly placed a strain on the British sustainment system as more US forces moved into Tunisia, so NATOUSA added the Eastern BS to the communications zone on February 13, 1943. Transportation difficulties forced leaders to recognize that the 250 miles from Oran to Algiers was too great a distance to provide logistics support to US units operating in the British First Army area. The Allies had insufficient trucks, the rail line was limited, and the road network consisted of dirt roads that frequently washed out. However, the North African coastline provided an opportunity to reduce distance from the ports to the combat forces by off-loading ships farther to the east for those supplies destined for use by the Eastern Task Force. The threat of attack by enemy fighters increased the farther east the coasters went, but the benefit of cutting hundreds of miles off the ground route made this risk worth taking. The main ports used by the Eastern BS included Bougie, Philippeville, and Bône.[31]

The theater staffed the new base section by reducing the size of the other two base sections, which created some animosity, especially from Brigadier General Wilson of the Atlantic BS. Brigadier General Larkin and the Mediterranean BS fully supported the levy of personnel and provided service personnel to the Eastern BS as required, but Wilson held out for as long as he could and gave people up only when forced.[32]

Each of the base section commanders had a vast area of responsibility and myriad responsibilities, including port operations, care of roads and rail lines, facility construction and management, lend-lease transactions, and civil affairs, to name but a few. Each commander was capable, but each also had his own education and experiences to draw from.

Arthur R. Wilson, commander of the Atlantic BS, had started his army career as an artilleryman in 1920. After completing the Command and General Staff course and War College, he became an army liaison officer with the Works Progress Administration from 1937 to 1939, followed by an assignment as chief of the Support Division of the Federal Works Agency from 1940 to 1942. These positions provided valuable experience in areas where most military officers had limited knowledge, such as jobs, civilian infrastructure, and function of a civilian bureaucracy. All of this would greatly aid Wilson's performance of his duties throughout the war as one of the Mediterranean theater's senior support commanders. The deputy theater commander, Hughes, described Wilson as being "an able officer . . . imaginative, but a free-wheeler who [prefers to jump over intermediaries and deal directly with senior commanders]."[33]

Patton had a less-generous opinion of the Atlantic BS commander, writing that "Wilson is nuts. [He] is a back-biting fool with an inferiority complex who will not last long."[34] Considering that Wilson was commander of the base section supporting Patton's forces in French Morocco, this comment seems to indicate a less-than-healthy relationship between the two men. Patton was ultimately wrong—Wilson did survive, but he did not achieve the same level of recognition as Thomas Larkin.

Thomas Larkin, commander of the Mediterranean BS, had received a commission in 1915 from West Point and entered the Army Corps of Engineers. He had a wealth of experience prior to World War II. He was part of the army's Punitive Expedition in Mexico in 1916, and he served with the Allied Expeditionary Force in World War I, where he received a silver star for the building of a bridge across the Marne River under heavy enemy fire. Larkin had a great deal of civil engineering experience as well. From 1929 to 1933, he was the district engineer

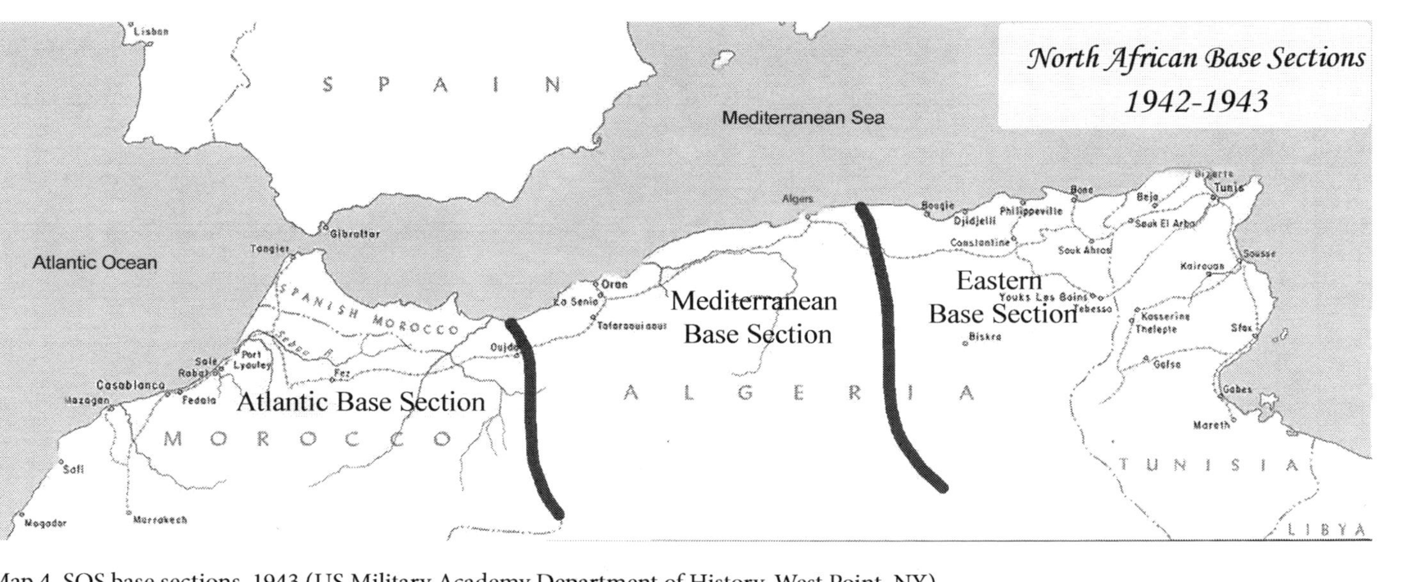

Map 4. SOS base sections, 1943 (US Military Academy Department of History, West Point, NY)

Figure 4. Brigadier General Arthur Wilson (US Army Center of Military History, Washington, DC)

for flood control at Vicksburg, Mississippi; from 1933 to 1936, he headed construction of the Fort Peck Dam at Fort Peck, Montana; and he was the officer in charge of the third locks project at the Panama Canal from 1939 to 1942.[35]

Similar to Wilson's experience in civilian institutions, Larkin's civil engineering background had a direct connection with the demands of operating a theater-support organization. The ability to leverage and improve civilian infrastructure was a large part of a base section's mission. However, unlike Wilson, Larkin was a personable officer who easily worked with others. This combination of civil engineering experience and affable personality led to Larkin's increasing levels of command and promotions throughout and after the war.

Figure 5. Major General Thomas Larkin (US Army Center of Military History, Washington, DC)

Arthur Pence, commander of the Eastern BS, had graduated from West Point in 1918 and, like Larkin, served as an officer in the Corps of Engineers. Pence did not have as much experience as Wilson or Larkin, but he did have a tour in the Philippines from 1926 to 1928 and was an instructor at the Engineer School at the start of the war. Pence was a proponent of getting the mission done in any way possible. By the use of both general policy and verbal guidance, he gave his men orders to "get the job done. Use all possible means. Improvise if the standard means and channels are inadequate. Keep all wheels rolling. . . . It is the results that count. . . . We must do everything possible ourselves without counting on help from the rear."[36]

Figure 6. Brigadier General Arthur Pence (US Army Center of Military History, Washington, DC)

AFHQ's direct control of the base sections ended on February 15, 1943, with the establishment of a new headquarters. SOS NATOUSA was an intermediate level of command designed to fall between the theater and the base sections. The SOS had the responsibility to direct and overwatch all sustainment activities for the entire North African theater. Larkin received a promotion to major general and left the Mediterranean BS to command the new organization. Colonel Edmond Leavey assumed command of the Mediterranean BS.[37]

The activation of the SOS NATOUSA was meant to simplify the management of theater administrative support; however, there was still disagreement regarding its exact functions and responsibilities. Major General Hughes, the deputy theater commander and commander of the communications zone, felt that although the three base sections operated under the oversight of the SOS for supply operations, they actually belonged to and reported to the theater. In this view, the SOS served not in a command role but rather in a position of management. Hughes felt that the AFHQ G4 should make policy and then pass it through either himself or Major General Larkin for execution.[38]

Tom Larkin held a slightly different interpretation. In his view, AFHQ intended that the SOS should have "command of all US Army supply and maintenance activities in the theater." As such, the base sections should report direct to the SOS and not to AFHQ. Larkin felt that he had command and control of "the operation of installations and the employment of personnel and troops in the communications zone." The SOS eventually stationed a liaison officer at AFHQ in an attempt to ease coordination; however, tensions between the two headquarters lasted well into 1944.[39]

The AFHQ G4 in the end fell into a role of developing logistics policy, setting priorities, providing logistical advice to Eisenhower, and conducting long-range planning. The G4 also coordinated common supply support for the air forces and adjusted levels of support between the ground and air forces for items with limited supply. While the G4 dealt with policy, the SOS handled the detailed planning of immediate support operations, the management of supplies, and the execution of all sustainment actions. In essence, the G4 focused on how sustainment actions should occur, and the SOS dealt with getting these actions accomplished. What remained unsettled was who had command authority over the rear area and the SOS. This same debate occurred in the European Theater of Operations during 1944, indicating that these types of command-and-control issues were common in theater-level operations.[40]

Overall, the SOS was responsible for providing all of the administrative support for US forces in the Mediterranean theater, which included supplies, transportation, medical care, signal support, civil engineering, and maintenance support. However, to do this it had to operate ports, build and run installations, build POW camps and provide supplies to the POWs, build telegraph and telephone systems, operate the rail system across North Africa, and maintain the roads, bridges, and rail lines.

The base sections established a wide range of capabilities to take care of the deployed force and to avoid returning equipment back to the United States for repair. Civilian shoe repair shops augmented military repair units, repairing some 400 pairs of shoes per day for one base section. Each base section contained dumps, warehouses, bakeries, sales stores, offices, ports, civilian and military hospitals, fuel-storage facilities, maintenance shops, lumber yards, map depots, wineries, equipment-assembly plants, motor pools, graveyards, rifle ranges, personnel-replacement depots, and rest camps.

The single rail line along the North African coast presented challenges of coordination between the Allies and the local French authorities. Compounding the problem was the initial lack of a central NATOUSA transportation section. Eisenhower described the challenges of the North African rail network in a letter to Marshall: with a rail system that could handle only nine trains per day, three of which were dedicated to nonmilitary uses, "the logistics situation is one to make a ritualist in warfare go a bit hysterical."[41]

On one occasion, the Eastern BS required a resupply of French francs to conduct purchases from local suppliers and venders. The SOS arranged with the Atlantic BS and the Mediterranean BS a shipment of one million francs under the plan to disguise the money in a piece of laundry equipment and to provide only a small guard. At some point in the journey, both the lone guard and the laundry machine were lost. An army laundry and bath unit later contacted the Atlantic BS and stated that upon opening a replacement laundry machine, they found a box full of French currency and asked for guidance on what to do with the newfound wealth. The base section eventually recovered all of the money except for a thousand francs. However, a better means of coordinating movements among the different support organizations could have easily prevented this entire incident from occurring.[42]

As the Mediterranean theater began to take shape, one thing became increasingly clear: the needs of the large Allied combined and joint force were outpacing logistics capability. Tom Larkin sent a clear message to AFHQ: the SOS needed adequate resourcing, or future operations and the lives of thousands of soldiers would be at risk. This message put both AFHQ and the US War Department on notice. Key positions within the SOS needed filling by qualified personnel, or there could be theater-wide and even strategic repercussions. To make matters worse, French forces had changed allegiances following the invasion and needed complete outfitting, thus placing new and unforeseen demands on the support system. As in World War I, requirements were outpacing capacity.[43]

The Battle for People

AFHQ clearly understood the need for additional personnel within the SOS, and by April 23, 1943, a cable was sent to the War Department requesting approval for an SOS headquarters of 594 personnel—double the size of the original authorization. By June, the fill of personnel still lagged behind authorizations. The theater supported the requested increases, but the reality of having too few trained service personnel in the army forced the SOS to be innovative and flexible and to make due as best it could. Some confusion remained; the SOS tried to create a casual detachment of 200 officers and 600 enlisted men, but Hughes canceled the move because AFHQ had not approved the proposal.[44] Larkin and his SOS were simply another demand amid a growing worldwide need for service forces as the different theaters continued to expand their level of operations. Until the fill of service units improved, North African base section commanders were simply unable to do everything they needed to do.[45]

Requirements for the air forces placed tremendous stress on the SOS. For example, the III Air Service Area Command on a daily basis required 250 tons of fuel and 50 tons of bombs and ammunition at the Tunis–Bizerte area, with an additional 450 tons of fuel and 250 tons of bombs and ammunition at Cap Bon. In addition, the Enfideville–Kairouan area needed 350 tons of fuel and 250 tons of bombs and ammunition. In other words, 1,050 tons of fuel and 550 tons of munitions were needed *every day* for this one air service area alone.[46]

Sustainment for the air corps and the airfields quickly became a major endeavor for the SOS. The rains that began in December 1942 rapidly turned airfields into mud fields. Broken stone laid on top of dirt had initially served as a serviceable runway during November, but after the rains arrived, the stone simply sank into the mud. Steel matting was the only appropriate material that could resurrect the runways and taxiways; however, a single runway required 2,000 tons of matting, which equaled two days' worth of the entire rail capacity for the forward area. This need for matting occurred at the same time that AFHQ was working to resupply the frontline forces, which was an even higher priority. The inability to move matériel forward to improve the airfields played a large role in Eisenhower's decision to abandon his hopes of an early seizure of Tunisia.[47]

Rearming the French—Three Hundred Thousand More Mouths

One development that provided a strategic boon to Eisenhower, but with a sizable logistic implication, was the agreement to rearm and outfit the Free French forces of North Africa. Under the terms of the agreements coordinated between the different commanders on the ground, French commanders retained command of their forces but agreed "to operate in a spirit of consultation and cooperation with Allied forces."[48]

The United States used the proposal of rearming French forces as a means to help convince French officers to join up with the Allied cause. By 1941, the Free French had established a national committee in London under the presidency of

General Charles de Gaulle. The British provided the initial equipment to some 30,000 Free French forces outside of Vichy control, using items supplied by the United States through the Lend-Lease Act of 1941. The connection between supplier and user had come full cycle—in 1917, the United States needed French help to outfit American divisions; twenty-five years later the relationship reversed.[49]

The capitulation of the French in North Africa dramatically added to the numbers of Free French forces. Through the agreement signed by Admiral Darlan and Lieutenant General Clark on November 22, the French agreed to join the Allied cause. This agreement, though, came with a cost because the French needed large amounts of training, equipment, and supplies. Despite these problems, they represented a large pool of manpower that Eisenhower could not ignore.

The French provided a sizeable addition to the Allied military strength in North Africa, stating that they could contribute up to eight infantry and two armored divisions within a month after the Allied landings. As French North Africa came under Allied control, the number of French forces increased the Allied militaries by 197,000 troops, and by November 22, 1942, the French had more than 300,000 men under arms, including an infantry division in Tunisia; three infantry divisions plus a light mechanized brigade in Algeria; and two infantry divisions, a light mechanized brigade, and 5,000 goum forces in French Morocco. In early 1943, the French forces represented available labor but needed an infusion of supplies and equipment to make them effective in combat. The French had lost half of their aircraft, tanks, and other heavy equipment. Many units had outdated equipment. The French soldiers needed everything from underwear to uniforms to rifles. After the force was reequipped, it also had to be sustained with ammunition, fuel, food, repair parts, and other like items. In addition, the French forces needed an overhaul of their entire sustainment system.[50]

Agreements reached at the Casablanca Conference of January 1943 stipulated that Eisenhower was the "final approving authority for French requisitions and that any rearmed French forces were to be given missions under the Allied Commander-in Chief's direction." This stipulation provided Eisenhower with a tremendous advantage. The Allies based the French–Allies relationship on cooperation, but if the French wished to enact a different strategy than the one the Allies proposed, all Eisenhower had to do was to threaten to cut off supplies for French forces. This very situation played itself out in December 1943, when the French Committee of National Defense declined to send a rearmed unit, the 9th Colonial Infantry, to the Italian front. After Eisenhower issued a telegram threatening to cut off the rearmament program, the Committee of National Defense reconsidered its position and agreed to send the 9th Colonial to Italy as AFHQ had requested.[51]

The rearming of the French was not just limited to ground forces. Speaking before Roosevelt and Churchill at the Casablanca Conference in January, the senior French commander, General Henri Honoré Giraud, asked to rearm the French air force as well—including fifty fighter squadrons, thirty light-bomber squadrons, and other transport squadrons. This totaled more than 1,000 aircraft. What at that time remained of the French navy sailed to the United States for overhaul and reequipping.[52]

In supporting French rearmament, the Allies gained a sizeable fighting force, one that was knowledgeable in local customs and terrain. However, the rearmed French forces also had a tremendous impact on the mission of the theater-sustainment units. The base sections received the mission to reequip the French forces using US equipment. During April to May 1943, the Atlantic BS equipped a French armored regiment, one tank destroyer battalion, an armored reconnaissance battalion, and eight antiaircraft battalions with all the necessary equipment to function as a fighting force, which included 1,067 vehicles, 586 trailers, 237 half-tracks, 214 tanks, 39 tank destroyers, and 111 antiaircraft guns. In addition, the base sections had to provide instructors to train French forces on the new equipment. To complicate matters, most of the French forces were Muslim, meaning that base section quartermasters needed specialized systems for rations distribution, wherein pork products were sent to units other than the French.[53]

The French would prove to play an important role in the Mediterranean theater through their force contributions, but the benefit was not without cost. To use the Free French, the United States had to provide most of the equipment and supplies needed to sustain these forces in combat. By the war's end, it had allocated 10,731 machine guns, 1,406 tanks, 27,176 trucks, and 1,417 aircraft to French use. This was a tremendous amount of equipment, especially considering the worldwide demand for such items. The Pacific theater and the Soviet Union were likewise clamoring for equipment, and any item provided to the French was one less available for some other requirement. Once the French units were equipped, the base sections kept them supplied with everything from food to fuel to ammunition and repair parts. Regardless of the cost, however, equipping the French provided at least an additional eight divisions to the Allies—divisions they otherwise would not have had.[54]

Although the decision whether to equip the Free French forces or not may seem simple, Eisenhower faced a complex and volatile situation with the Free French, which significantly complicated his job as a theater commander. French lines of command and control were unclear. Admiral Darlan was assassinated on December 24, 1942, and General de Gaulle was working to be recognized as the leader of all Free French. French politics, international domestic relations, and guidance from Roosevelt and Churchill combined to make the situation with the Free French anything but straightforward.

American ordnance units assumed responsibility for the rearmament of the French. Plans called for French forces to assemble the lend-lease vehicles arriving from the States, but the French lacked the equipment, personnel, and experience necessary to accomplish the task. Schools had to be set up to train the French on the unfamiliar equipment. The French had good attitudes regarding fighting but failed to take care of equipment. The increased wear and tear of US equipment issued to the French placed an additional burden on American maintenance units. Moreover, manuals and stock records required translation into French, and requisitions in French had to be translated into English.[55]

As the theater grew, the war became a war of attrition, which meant that it became a war of resources. The Americans were trying to find a workable system of command for support forces, while British forces opted for a simpler approach.

The British Approach to Support

The British support system was a bit more streamlined compared to the American system, and the British had more experience working in North Africa. Major General Gale wore two hats—he served as the chief administrative officer within AFHQ while also providing guidance to the British Number 1 Line of Communication. In the latter role, he performed duties similar to those covered by Major General Hughes, the US deputy theater commander. The major difference, however, was that Gale's directions to the Number 1 Line of Communication were treated as orders, whereas Hughes's directions to the SOS did not carry quite the same weight. Bedell Smith noticed the difference and worked for at least six months to encourage Eisenhower to adopt the British system, an argument Smith never quite won. Despite the differences, Hughes and Gale worked in the same headquarters, proving opportunities to synchronize US and British support even though the two nations handled the execution of support independently. All the combatants had to deal with the same challenges of austere infrastructure, difficult terrain, weather, and long distances from home ports.

In the Eastern Task Force area, the main forward supply line ran through Tébessa eastward to Souk el Arba. Up to nine trains traveled the route daily, with one allocated for civilian materials, two with coal, and six with military supplies. The port at Bône became a major part of the sustainment system for forces in northern Tunisia. As in the Western and Center Task Force areas, British-controlled ports had challenges keeping up with demand, and matériel piled up due to the inability to effectively move supplies inland. The same limited road and rail network seen in the western part of the theater plagued the eastern part as well.[56]

As the theater matured, the British placed headquarters of the Number 1 Line of Communication at Sétif with supporting service commands and bases located at Algiers, Bougie, Philippeville, Bône, Constantine, and Souk Ahras. The maintenance facility for the British Churchill tanks was located at Bône. National-level support was worked through the British Supply Agency and the British Middle East Command, located in Egypt. As with the Americans, the British developed a comprehensive theater-support operation that executed the priorities and directions developed by AFHQ.[57]

Logistics on the Other Side of the Hill

The German supply situation was dire even before the Torch landings. The Afrika Korps commander, Field Marshal Erwin Rommel, complained to his superiors that in the first eight months of 1942 German desert forces had received only 40 percent of the 120,000 tons of requested supplies. Continuous operations over the open desert ruined German vehicles, and there were not enough engineer units or resources to build more roads.[58]

Despite the shortages in supplies, Rommel did have success in 1942 and early 1943. Generalmajor Hans-Henning von Holtzendorff, commander of the 104th Rifle Regiment in North Africa, attributed Rommel's early desert victories to Rommel's personality, bold decision making, strong nerves, ability to find new ways of fighting, and utilization of the element of surprise and improvisation. Holtzendorff noted that interference of the sea line of communication was a potential problem for both Allies and Axis alike, although the British did have a longer but safer route available around the Cape of Good Hope.[59]

Supply of North Africa was the responsibility of Comando Supremo, the Italian high command. Support requests for German and Italian divisions flowed from North Africa to Italy, where the Italians filled what they could. In addition, the German High Command, the Oberkommando der Wehrmacht, shipped matériel and equipment from Germany to Italy, where it could be fed into the Italian distribution system. Supplies went by rail to Italian ports, where German or Italian ships transported them to ports in North Africa under Axis control and away from Allied air cover. Daily movement of war supplies for German forces averaged forty-eight trains per day between the Reich and Rome. Once in North Africa, the unloading and inland distribution were the responsibility of the supply and administration officer, initially of the Afrika Korps and later of Panzergruppe Afrika (later redesignated Panzer Army of Africa and then German-Italian Panzer Army). Luftwaffe units assigned to North Africa received all rations, pay, office supplies, lodging supplies, and fuel from the Wehrmacht, the unified armed forces of Nazi Germany.[60]

In the fall of 1942, the Axis army in North Africa was in the middle of a long retreat, heading west, back from El Alamein, with the British Eighth Army in pursuit. Less than a third of German tanks were serviceable, and British aircraft were effectively targeting the fuel tankers destined for Rommel's North African ports. Most of the Italian infantry were dismounted, meaning that they needed external resources to move about the battlefield. The situation became only direr as the operation went on.[61]

Rommel and the Axis force of North Africa faced four problems. First, Hitler considered North Africa a secondary front, so it received fewer resources from the German High Command. Second, Malta served as an effective base of operations for the British air and sea services in the Mediterranean, allowing them to target the Axis supply convoys steaming out of Italy or southern Europe. Third, the Germans did not have a single overarching joint sustainment organization within North Africa to handle the organization of the supply effort or to centrally manage limited resources. Each military service took care of its own needs, and there was little prioritization or theater-level control. This meant that the German army might divert a shipment of fuel to the front lines even though that fuel could achieve a greater good with the Luftwaffe. Conversely, the Luftwaffe might divert fuel to its airfields when the army really needed it at the front. Finally, the Germans did not have the truck capacity needed to sustain an army over extended distances for long periods.

By January 1944, Rommel noted that the shortage of supply trucks and the long distances meant that he could receive only 50 tons of supplies per day rather than the 400 tons that he needed. "The result was to be seen in deficiencies of every possible kind." Perhaps part of the problem was also the lack of importance the Germans seem to have assigned to logistics staff officers. In the US theater, Eisenhower had a major general in charge of the communication zone, a major general heading the SOS, and a brigadier general serving as the US theater G4. In contrast, Rommel's supply officer was a Major Otto.[62]

Field Marshal Albert Kesselring, the Axis commander in chief and Rommel's immediate superior, believed that the protection and organization of the overseas lines of communication was of decisive importance and that the Italian navy and its weak air forces had failed to secure those lines. In fact, Kesselring later labeled the failure to protect the sea lanes between Italy and its colonies and the deterioration of the German Supply Service in Tunisia as two crucial Axis mistakes in the North African campaign.[63]

A concerted Allied attack on the Axis lines of communication achieved the desired effect. Kesselring confirmed that the interdiction of the sea lines by Allied air, marine, and submarine forces kept supplies from arriving in North Africa. A similar attack on German ground traffic within North Africa made the supply situation "intolerable." The Germans had enough problems supplying Rommel when he was facing only the British Eighth Army; the addition of another Allied force to the west made North Africa unwinnable for the Axis.[64]

In his analysis of Rommel's campaign, the historian Martin van Creveld concludes that the Axis problem in North Africa was multifold: the need to secure sea lanes, limited Libyan port capacity, and the associated distance between the ports and the front. Rommel needed 100,000 tons of support matériel each month, but his closest supply port (Tobruk) had a capacity of only 20,000 tons and was hundreds of miles behind the front lines.[65] Benghazi and Tripoli had greater throughput tonnages but were 800 to 1,300 miles from the front. Ports closer to the front often fell within range of the Royal Air Force, preventing their effective use. Creveld argues that Rommel received sufficient quantities of fuel between July to October 1942 but lacked the means to effectively distribute it. He faced a system problem: convincing the Italians to provide sufficient supplies and shipping them, getting the convoys past Malta, finding ports near the front with sufficient throughput, and then having the means to prioritize the distribution of the matériel and transport it to the right place at the right time. Rommel had challenges with all of these parts of the system.[66]

In December 1942, both sides had limited supplies and transportation. In the race for Tunisia, the side that could more effectively and efficiently produce and distribute the means to fight had the clear advantage. On December 20, the question of supplying Africa was the subject of a conference at Hitler's headquarters near Rasternberg. The Italians agreed to move supplies forward and to convert three destroyers into supply ships; neither happened. On the Allied side, the SOS worked to continue setting up the theater. For Allied and Axis alike, the support effort might not win the war, but the lack of support could cause one side or the other to lose it.[67]

By April 1943, the North African theater was set. AFHQ was the Allied theater headquarters, and NATOUSA served as the US theater headquarters. The SOS managed work in the communications zone. The British Number 1 Line of Communication handled support of all forces under British command. The base sections were maturing with a growing capacity, while the service units within the base sections found themselves bringing order out of chaos via available military and civilian resources and infrastructure. Eisenhower was now ready to finish the fight in North Africa, and the communications zone was set to support the operation.

3

The Fall of Tunisia

The battle is fought and decided by the quartermasters before the shooting begins.
—Field Marshal Erwin Rommel

With the missed opportunity to grab Tunisia in December 1942, Eisenhower faced the need to establish a capable theater that could drive Axis forces off the African continent, and the early months of 1943 provided just the opportunity. Hitler wanted to support his field commander in North Africa, Field Marshal Erwin Rommel, but competing priorities and recent Russian success along the Eastern Front limited Axis options. The Allies continued their buildup of theater infrastructure. As a result, the North African communications zone of 1943 was much more capable and agile than its European predecessor in 1918. North Africa now possessed improved transportation infrastructure, better management of transportation resources, more mission-ready trucks, and numerous supply dumps. Allied aircraft routinely attacked Axis shipping in the western Mediterranean. The Allies were ready to finish the fight in Tunisia and to begin the next step toward victory in Europe.

In early 1943, however, the Axis forces of North Africa were not yet ready to hand the Allies an easy victory. Rommel still had a chance to strike deep into Allied territory and cause real harm to the Torch forces before turning his attention to the East and dealing with General Bernard Montgomery's Eighth Army. The battle for Tunisia was not a fait accompli.

Early 1943: Advantage Rommel

The failure to capture Tunisia in 1942 forced a reexamination of Allied strategy. One of the items agreed to at the Casablanca Conference of January 1943 was an invasion of Sicily in the summer of 1943. This meant that the Allies had to finish the fight in Tunisia during the spring of 1943 in order to meet the deadlines associated with preparing for this follow-on invasion. Unfortunately, winter rains needed to end before AFHQ could launch a full-scale offensive, and the rainy season traditionally lasted until the end of March.[1]

The rains, terrain, and enemy situation outlined the parameters of the Allied plan. Rommel's force had retreated up into Tunisia. The British Eighth Army extended west, out of Tripoli. Eisenhower's forces would have to hold the western flank until conditions improved enough to support an attack in the spring of 1943.

By January, Rommel commanded approximately 100,000 Axis forces: 74,000 Germans, and 26,000 Italians. The Italians were dismounted infantry and needed German transport to keep up with the German mechanized units. German units were at only about half of their authorized strength, and only one-third of their equipment was operational. The Allies, for their part, were severely under-manned, with three corps defending a 250-mile front along the Tunisian western dorsal passes.[2]

Fighting in January occurred generally along the eastern dorsal passes of Tunisia. Allied units possessed better logistics but were relatively inexperienced. German units had experience and shorter lines of communication but lacked the units, cooperation, supplies, and equipment needed to achieve a breakthrough.

Eisenhower repositioned II Corps to the east in an attempt to push through the German lines and reach the port town of Sfax, thus splitting Rommel's force in half. The Eastern BS pushed its supply depots forward by using the rail line from Constantine to Bône and then to Souk Ahras at a rate of 250 tons per day. Supply points extended as far east as Kasserine. However, a German attack against Faid Pass on January 30 convinced Eisenhower to call off the Allied attack and to put II Corps on a defensive setting instead.[3]

In February, Rommel tried once again to break through the western Allied lines. Fearing that the Americans might again try to split the Axis forces, he gained approval to attack the US formation on the west and then swing his forces to the east to deal with Montgomery. The plan called for General Hans-Jürgen von Arnim's Panzer Army to strike west from northern Tunisia, while the Afrika Korps converged from the south, forming a pincer attack in the direction of Kasserine.

The attack commenced on February 14, led by the 21st Panzer Division with its eighty-five tanks streaming from the Maizila Pass. Simultaneously, the 10th Panzer moved through the Faid Pass. Three days later, the 21st and 10th Panzer Divisions had seized their initial objectives. The Fifth Panzer Army character-ized its supply situation as acceptable. Rommel was ready to push on through Kasserine Pass in an attempt to seize the port of Bône, a move that would out-flank Eisenhower's forces. The attack began on February 19.[4]

The Axis attack made rapid advances against inexperienced and ill-positioned US and British units. US service units destroyed ammunition, gaso-line, and other supplies near Sbeitla in fear of the German advance. The US gar-rison evacuated the town of Gafsa. The drive in Tébessa convinced II Corps to evacuate its large supply dumps. Units set fire to everything they could not move. The II Corps railhead at Ferianna held approximately 55,000 gallons of fuel—fuel that the service units could not evacuate but would be a prize for the German Panzers. Allied units in the area filled their vehicles and then the quartermasters fired tracer rounds into the remaining drums and fuel cans, igniting the dump. Tébessa became the fallback position for II Corps support units and the main support base for the battle.[5]

Whether Rommel realized it or not, Tébessa represented the one objective that could provide his divisions with the resources needed to sustain their drive to Bône. With the US pullback from forward positions on February 16, Tébessa now held more than half a million gallons of gasoline and a million rations. If Rommel overran Tébessa, he could seize the supplies needed to keep his armies on the offense.[6]

A lack of support combined with poor road conditions slowed the Axis attack, however. The Germans controlled Kasserine Pass by the night of February 19, but delays allowed II Corps and the British to reinforce its positions. By February 22, Allied units had reinforced themselves to the point that Rommel felt there was little chance of further success. The Germans had only a few days' worth of food and ammunition remaining and little reserve fuel. On February 23, Rommel pulled his units back to the east, giving back Kasserine.[7]

The battle was costly for Eisenhower. Besides the human casualties, II Corps also lost 183 tanks, 194 half-tracks, 208 pieces of artillery, and 512 trucks. Units needed replacement equipment before the Allies could consider a large attack. In addition, some items, such as the Sherman tank, needed immediate updating to deal with superior German firepower. What is notable, however, is the fact that even though the American forces lost more armored vehicles than the Germans had attacked with in total, these losses were quickly replaced. The American logistical system was finding its stride.[8]

Following the battle, Hitler tried to fix the German command-and-control problems by placing all German ground units under Rommel's command, with the title Army Group Afrika. However, Rommel's time in Africa was limited. Following Kasserine, the German commander had increasing difficulty with his heart, nerves, and rheumatism. On February 26, Rommel wrote to his wife complaining of his health problems and mentioned that the conditions for victory did not exist in North Africa. "Everything depends on supplies," Rommel wrote. Ultimately, neither the German supply situation nor Rommel's health would improve. On March 10, Hitler recalled Rommel to Germany for health reasons and passed command of Army Group Afrika to General von Arnim.[9]

Allied interdiction of Axis supply convoys in the Mediterranean was having an impact. In the spring of 1943, Axis planners assumed that large ships could be used only one time for a supply run to Africa before becoming lost. The total monthly requirements assessed by German commanders totaled 150,000 metric tons, but limited shipping space reduced the ability to meet this need. Efforts to enlist small boats fell short, and a follow-on project to construct concrete cargo vessels failed due to lack of engines. Axis prospects appeared increasingly bleak.[10]

Both sides used April to consolidate their positions, rest the troops, and prepare for the next fight. The Allies needed to reequip themselves and reestablish the supply dumps evacuated during the German advance. Salvage crews from the Eastern BS gathered up debris from the battlefields and trucked it off to depots for sorting and repair. Engineers worked to reestablish damaged rail lines and roads. By April, the Eastern BS had restocked the theater, and Patton was now in command of II Corps after the firing of Major General Lloyd Fredendall. Moses brought the Ten Commandments to the Israelites to restore discipline and order; Patton relied on army traditions and regulations. The Allies were preparing for the final drive through Tunisia.[11]

Regaining the Initiative

As II Corps moved east, the Eastern BS's boundary was moved east as well. This followed established doctrine—as the combat zone shifted forward, the leading boundary of the communications zone also moved up, which ensured that the combat forces could maintain a focus on the enemy without having to dedicate additional resources to maintain a lengthening supply line. However, someone had to deal with the lengthening lines of communication, and this job fell to the service units in the communications zone. By April 8, 1943, the Eastern BS had moved its boundary as far east as Tébessa. Advance depots became general depots. Engineers worked to improve road and rail lines to support the increasing usage of the limited infrastructure. The base sections used all available seaports along the northern coastline to support the flow of matériel into the theater.[12]

The final phase of the North African campaign started in April 1943. The combined pressure from the British Eighth Army in the East and from the British First Army and US II Corps in the West forced von Arnim to collapse the German defense in southern Tunisia and concentrate his forces in an arc in northeastern Tunisia, which allowed Axis forces an opportunity to concentrate their units and shorten their lines of communication. General Sir Harold Alexander of Great Britain, Eisenhower's commander for ground operations, devised a plan to stage II Corps along the northern part of the front. Montgomery's Eighth Army was to attack from the south. The main effort would come from the British First Army, positioned south of II Corps.

In response to the developing situation, the Allies reorganized their lines in Tunisia in mid-April 1943. The British shifted forces from the Eighth Army on the coastal plain to the zone of British IX Corps east of Le Kef. The US II Corps moved 140 miles to the north to occupy an area in northern Tunisia east of Bédja. This move, which occurred in only four days, from April 14 to April 18, forced a complete shifting of the eastern US supply line. There were competing demands on the limited road and rail networks in the region because the British were hesitant to allocate more than 250 tons per day out of the available rail capacity. A deal between the SOS and the British allocated an additional 250 tons per day between Souk Ahras and Bédja in exchange for British use of an equal tonnage on the American narrow-gauge track between Ouled Rahmoun and Tébessa. Arrangements like this one supported the needs of both armies and allowed the Eastern BS to shift its line of communication to the front while also building a stock of supplies for the upcoming fight.[13]

To support operations in Tunisia, the Atlantic BS and Mediterranean BS acted as intermediate-level bases, providing bulk supplies to the Eastern BS, which directly supplied the combat zone. The Mediterranean BS established an advance ordnance depot at Constantine, capable of supplying all US forces east of Algiers with fifteen days of supply. This push of supplies allowed the Eastern BS to focus its efforts forward toward II Corps, with the main supply route running from Constantine to Tébessa.

To meet the increased demand, the Eastern BS established supply depots at Tébessa, Ouled Rahmoun, Bône, Mateur, Tabarka, and Philippeville—with the one at Philippeville supporting more than 40,000 troops. Due to labor shortages, however, all depots operated with reduced staffs.[14]

For the attack across Tunisia, the Eastern BS established a stockage goal of twenty-one days of supplies and enough ammunition for six units of fire. Trucks from the Eastern BS delivered supplies to II Corps supply dumps, which were typically 15 to 30 miles behind the forward areas of the combat zone. German air attacks had subsided by this time, so food, fuel, and ammunition distribution generally occurred during daylight, and the dumps required no camouflage, only dispersion for security. Once supplies were at the dumps, II Corps truck units pushed supplies forward to the frontline units.[15]

Axis air forces were a shadow of their former selves. The Luftwaffe was operating at half strength, down to 178 aircraft, since January 1943, and the Italians had only 65 aircraft. In comparison, the Allies had 3,000 available planes. Allied interdiction of sea lanes had drastically reduced levels of resupply, especially fuel. Between March and May, Axis forces lost 230 ships and 42 percent of fuel destined for Africa.[16]

The British First Army had operated in Tunisia since the landings of Torch, so it was closer to British ports and a developed support system. In contrast, the British Eighth Army, staging from Egypt, faced an ever-growing line of communication. Pursuing Rommel since the battle of El Alamein in November 1942, the Eighth Army had to deal with a line of communication that extended itself day after day. The capture of ports along the African coastline, such as Tobruk, helped shorten the overall distance that supplies traveled by road but extended British sea-shipping distances. In addition, as the army moved forward, the British support bases moved as well. The arrival of General Montgomery's forces into central Tunisia in March 1943 allowed the Allies to reorganize their lines of communication and begin supporting Montgomery from the west.

In April 1943, the Eastern BS opened a supply line between Ouled Rahmoun and Tébessa, providing a new line of communication to Montgomery's Eighth Army, which up until then had been receiving its supplies from as far away to the east as Tobruk and Egypt. The rapid movement into Tunisia and the contraction of the front had caused major supply challenges for the 18th Army Group, so the diversion of US supplies provided needed relief. The new line of supply was just 100 miles long. As the British pressed north along the Tunisian coastline and the Allies liberated the Tunisian ports of Gabes and Sfax from the Germans, the supply line reversed itself. The Allies were able to off-load ships at these ports and to truck supplies overland back to the base section depots, thus providing shorter routes from the ports to the forces and adding more port capacity to the SOS.[17]

Upon recapturing the airfield at Gafsa, II Corps turned the facility into a forward supply dump for Montgomery's use as well, which allowed the Allies to stockpile resources in anticipation of the Eighth Army's northward advance. The plan worked, and by April 15 the British Eighth Army had moved north through the Mareth position and stood before the Axis defensive Enfidaville line.[18]

The distribution of fuel within North Africa occurred through a combination of pipelines, rail, and trucks. Pipeline and rail provided the most efficient means of movement, but at some point quartermaster units had to transfer the fuel into smaller containers, put these containers onto trucks, and then transport them to the front lines. To complete the system, empty fuel cans needed to be collected and then returned to the support areas for refilling.

The 2004th Engineer Pipeline constructed a pipeline from Philippeville to Ouled Rahmoun as well as a bulk-storage tank farm on April 6, 1943. This pipeline had the capacity of moving 500 tons of fuel per day, the equivalent of the capacity of 200 two-and-a-half-ton cargo trucks. The pipeline proved to be so successful that a second pipeline was started a week later from Bône to Souk el Arba. Together, these two pipelines moved 1,200 tons of fuel per day, freeing up 480 cargo trucks for transporting other matériel. During the month of April alone, the Eastern BS provided more than 7,800,000 gallons of fuel to the US II Corps and Air Corps.[19]

Sabotage was a constant problem because these pipelines were exposed and largely unprotected. Mounted Arabs with incendiary bullets often hit the pipeline, causing a spout of flaming fuel to leap upward. Roving guards failed to stop the sabotage. One day a reserve officer who had worked in civilian construction and knew the cultures of the French and Arabs suggested that the Eastern BS should contract out security of the pipeline to the French and levy fines if the pipeline were damaged. The French, in turn, would contract the Arab caliphs with the same provision of a fine for any damage. Pence supported the idea, so the Eastern BS made the proposal to the French. The French government agreed to the contract as long as they could execute it as they saw fit. Almost overnight the shooting and vandalism stopped, US guards were able to return to their normal jobs, and the fuel moved forward.[20]

Despite efforts to increase the number of trucks in North Africa, transportation shortages continued into March 1943. Base section planners had estimated that the rail line from Philippeville to Ouled Rahmoun could handle 1,500 tons per day, but they soon fell quickly behind due to the limited network. Following the German breakthrough at Kasserine Pass, the Allies had to work quickly to move ammunition and other essential supplies forward to support a counteroffensive against Axis forces. To meet the increased demand, the SOS formed new truck companies and leveraged resources from across the communications zone. A single truck regiment, the 46th Quartermaster, handled the control and administration of more than 6,000 troops. The SOS also formed a Movement Control Group to provide for operational control of all trucks. Engineer, armored, infantry, and artillery units provided the drivers.[21]

Challenges accompanied the formation of these new truck units. Soldiers were untrained on convoy operations, and their officers were untrained on leading truck units. There was an overall shortage of tools needed for vehicle maintenance, and even if the tools were available, there was a shortage of repair parts. One unit arrived in the area with no tools other than the few small kits that accompanied each vehicle.[22]

The Allied attack began in the British Eighth Army sector the night of April 19–20, 1943. The British initially broke through the Italian defensive line, but General Fritz Bayerlein, commander of German forces in the Italian First Army, quickly repositioned forces to plug the hole, stalling the British attack for four days. The following night, April 20–21, the Germans made their own attack against the British First Army and penetrated about five miles before stalling against the main defensive positions.[23]

The British First Army launched its attack on April 22, which quickly developed into a major battle, breaking through the German Afrika Korps's northwest flank. Divisions fought each other, incurring heavy losses and consuming large amounts of fuel and ammunition. By April 26, German supplies were running low. Limited amounts of fuel constrained the movement of German tanks and trucks. Axis units were short of the munitions needed to repel a strong Allied attack. By April 30, the lack of fuel immobilized the Axis force, but it kept on fighting and somehow held off Allied attacks in the critical sector of Djebel Bou Aoukaz.[24]

On April 23, II Corps, now led by Major General Omar Bradley, started its attack to the east in an attempt to move through the mountainous Djebel region and head toward Mateur and Bizerte. The terrain in the 9th Infantry Division's area was especially rough, with few roads and limited mobility. The SOS and the Eastern BS had built a sizeable line of communication forward from the ports, but the end of the chain for many combat battalions involved moving supplies by mule for several miles. II Corps made steady advances. A daily train from Bédja carried essential supplies forward from the supply dumps. At times, the corps resorted to the use of combat vehicles, such as tanks, to move artillery ammunition forward.[25]

The Allied advance continued into early May, pushing the front north and east. In Italy, Kesselring begged for supplies for North African operations, but the German High Command and the Italians ignored the requests. On May 3, the city of Mateur fell to II Corps. General von Arnim tried to reinforce Bizerte, but the 10th and 21st Panzer Divisions could not move because they were out of fuel. Axis planes and ferries landed 1,130 tons of ammunition and 180 tons of fuel on May 4, but it was too little and too late.[26] II Corps captured Bizerte on May 8, and the British First Army rolled into Tunis on May 9, splitting the Axis force. Von Arnim's southern flank collapsed on May 12, which led to the final surrender on May 13. The final transmission from the Afrika Korps read, "Ammunition shot off. Arms and Equipment destroyed. . . . D. A. K. [Deutsches Afrikakorps] has fought itself to the condition where it can fight no more." The Allies had finally gained victory in Tunisia and North Africa.[27]

Along with the Tunisian offensive in May 1943, the Allies became responsible for the housing and care of a large number of German and Italian POWs. The base sections had to transport, guard, feed, and care for these prisoners. By the time Tunisia fell, 275,000 Germans and Italians crowded the POW camps. To support the burgeoning prisoner population, base section units used captured German rations, and field kitchens helped offset demands on Allied rations. Empty wine vats became storage for potable water. Fortunately, morale among the prisoners was good, and they needed guides more than they needed guards.[28]

The POWs represented another demand on the supply system for resources, but they were also a possible source of labor. The US Army had an ongoing shortage of service forces within the force, and local Arab labor in North Africa was largely untrained and undependable and required close supervision. To take advantage of the situation, SOS NATOUSA instituted a program aimed at Italian POWs that offered compensation in exchange for their labor and support of the Allied war effort. Of the eventual 65,720 Italian POWs who remained within the Mediterranean theater throughout the war, 62,089 served in organized POW service units. This was the beginning of a program that by the summer of 1944 allowed SOS NATOUSA to augment or replace approximately 175,000 military support positions with Italian POWs and civilians.[29]

By May 13, the Tunisian campaign had ended. The campaign seemingly took a long time to complete, but in actuality only seven months had passed between the initial landings and the Axis surrender in Africa. In this short period, the Allies had landed a sizeable force on the African continent, established a theater of war, expanded their capabilities, and learned to adapt to the difficult environment. US forces alone had grown from 128,567 in November 1942 to 395,461 by April 1943—a threefold increase. In contrast, Axis forces lost more than 230,000 troops and much of the Italian fleet and had expended precious supplies that could have been used elsewhere. The Allies lost 770 aircraft, compared to the 2,000 lost by Axis forces.[30]

Enabling this victory was the establishment of a functioning communications zone. The base section commanders had a personal interest in caring for the forces they supported. SOS NATOUSA had oversight of the entire support structure and could move resources quickly as the circumstances demanded. Its responses to challenging situations, such as the need to increase ground transportation, proved that the SOS/base section doctrine was indeed flexible and could adapt as needed to meet the demand of the frontline units.

However, much also depended on the quality and timing of the decisions made by key leaders. Commanders in both the combat and communications zones had to understand the needs and limitations of the other in order for the theater to function. Eisenhower's decision to delay offensive operations in the winter of 1942 to allow the support structure to build was just one example of a decision that was unpopular but necessary and operationally correct. Ultimately, victory was assured—not just because of the maneuvering of the Allied force but also because of the capabilities provided by the sustainment system.

There was clearly room for Allied improvement. American forces were often outgunned; the Stuart and Grant tanks' main guns were too small and their armor insufficient. American generals from Eisenhower on down displayed leadership and decision-making skills that were less than inspiring. Many British generals looked down upon their American allies.[31]

After the war, von Arnim wrote that the failure of Axis forces in North Africa was caused by a breakdown in supply, with four main contributors. First, was lack of an effective combined command within the Mediterranean. Arnim writes, "One thing was clear—I had nothing to do with supply. This task belonged to the HQs personnel of Kesselring." Second was a hopeless inferiority of aerial forces. Third was the failure of Axis forces to interdict Allied air and sea supply lines. Last was the failure to take Malta.[32]

Interestingly, almost every German general interviewed after the war seemed to come back to Malta as a prime reason why the Axis failed in North Africa. Major Richard Feige, a supply and administrative officer within the Fifth Panzer Army, noted that following Operation Torch, Axis supply convoys were split up to reduce the chance of losses, and the Italians were forced to rely on smaller, faster craft instead of large cargo ships because the smaller craft could evade the Allied blockade. These factors led to a complete breakdown in the Axis supply and effectively immobilized the German–Italian Panzer Army after the withdrawal to Tripolitania. In an overseas campaign, sea lines of communication must be a prime concern of the highest levels of command.[33]

The window of opportunity for Axis victory in the Mediterranean was closing. Despite questionable decision making early in the campaign, Rommel recognized tactical prowess among American troops and understood that German superiority in weaponry would be short-lived once the US industrial base had time to react. Time was on the side of the Allies.[34]

The surrender of the Axis forces in Tunisia signaled an opportunity to reset Allied forces, plan for the next operation, and discuss the lessons of the past six months. There were many lessons, but whether the Allies would learn from their mistakes was an entirely different issue. During the Casablanca Conference in January 1943, the Allies had agreed to continue the campaign in the Mediterranean; now the invasion of Sicily was less than two months away. Sicily was a logical objective because it allowed Allied leaders to use forces and airfields that were mostly already in theater. Supplies for the invasion were already on the move from America and Great Britain. Grudgingly, the United States agreed to continue with the fight in the Mediterranean but didn't have much time to digest the lessons of North Africa.

Part II

Sicily and Italy

Journeyman

4

Operation Husky

Sicily

The main thing is to get the men and stores ashore. If they are there, they can be used.
—Commanding general, Seventh Army, report on the Sicilian campaign

The year 1943 proved to be a critical period of the war as momentum shifted to the efforts of the Allied nations. In February, Field Marshal Friedrich von Paulus and the German Sixth Army surrendered at Stalingrad, dealing Hitler a massive loss of men and a corresponding public-affairs disaster. After Stalingrad, Russian forces would continue to conduct a series of counterattacks to the west, with Stalin all the while demanding help from his allies.

February also saw US victory at the Battle of Guadalcanal, a major Allied win in the Pacific. Admiral Ernest King, commander of the overall US fleet and the chief of naval operations, made strong arguments for an increase in the allocation of units and resources for the Pacific. Among the items needed were ground and air service units as well as landing craft—the same resources needed across the Mediterranean and in England. The battle for the Pacific was frustrating for commanders such as Admiral King and General Douglas MacArthur. Not only were they trying to win the war against Japan, but they also had to win the war for resources back in Washington. With a Europe-first policy, the Pacific theater had to fight for everything it needed.[1]

Naval progress, though, was not just limited to the Pacific. In March, improved Allied antisubmarine tactics forced German admiral Karl Dönitz to withdraw a majority of German U-boats from the Atlantic because of heavy losses, which helped to ease the pressure on the shipping lanes and thus increased the delivery of supplies and matériel to Great Britain and the Mediterranean. However, the battle against the U-boats was still far from over.

The agreements that came from the Casablanca Conference envisioned the war unfolding along several major efforts. First, Eisenhower's forces would occupy Sicily in order to open the western Mediterranean to Allied shipping. Second, England would be the base for a large Allied strategic bombing offensive against Germany as well as serve as a staging base for a ground force capable of reentering the European continent. Third, the Allies would continue operations in the Pacific with an eye toward increasing the tempo against Japan once Germany was defeated. Churchill and Roosevelt approved of this plan while also maintaining that air reinforcements needed to go to China and supply convoys for Stalin.[2]

Roosevelt remained committed to a cross-channel invasion of France, although by the time of the Casablanca Conference it was clear that this could not occur in 1943. The United States had pulled too many units and supplies out of Great Britain for use in Operation Torch. The president reassured the prime minister that the United States would rebuild its base of operations in Great Britain for use either as a striking force for Operation Sledgehammer or as the nucleus of a larger invasion force for Operation Roundup. However, with the diversion of units and supplies to the Mediterranean, Operations Sledgehammer and Roundup—the invasion of France—could not occur before the spring of 1944. Instead, the Allies continued the march east across the Mediterranean, setting up bases of operations as they went.[3]

Although the fighting in Tunisia continued in January, Allied leaders understood the need to determine the follow-on strategy so planners and logisticians could begin the work of preparing for the next operation. On the final day of the Casablanca Conference, Roosevelt and Churchill agreed that to make continued use of the force in North Africa, the best option was to pursue the capture of Sicily in order to clear Mediterranean shipping routes and open a path directly toward the Italian Peninsula.

The Planning of Operation Husky

On February 2, Eisenhower issued a preliminary directive for his senior commanders to begin planning for the seizure of Sicily. Outline plans were due to the commander in chief by March 15, meaning that planners had less than six weeks to develop the basis for a campaign plan involving several divisions and the largest amphibious operation yet attempted. The commander in chief believed that there were two main tasks: the first to get sufficient troops and supplies ashore as quickly as possible, the second to secure ports for the further buildup of these forces. Unfortunately, Allied plans failed to place importance on the only objective that really mattered—seizing Messina and thus cutting off the Axis route of escape.[4]

Sicily is a large island of some 15,000 square miles. Much of the island's topography consists of rocky ridgelines and hills, and it has only a limited supply of freshwater. Movement was generally limited to existing road and rail networks, making cross-terrain movements and resupply efforts difficult, if not impossible.

Existing highways were in generally good shape, especially compared to the roads of North Africa. Standard-gauge rail lines circled the island, and auxiliary narrow-gauge lines connected interior towns. Rapid, large-scale military movements were possible, but they required sufficient rail and/or motorized vehicles.

Sicily has four large ports: Messina, Catania, Syracuse, and Palermo. Messina, on the northeastern corner of the island, served as the main Axis port. This port also had extensive air defenses and fell under the umbrella of Axis air coverage from mainland Italy. Catania, located midway along the island's east coast, also contained sizeable defenses. Any attempt to seize these locations would be extremely tough, if not suicidal.

This difficulty left planners with the options of Syracuse and Palermo. Syracuse, located on the southeastern corner of Sicily, did not contain as many Axis units and was within range of Allied fighters and medium bombers originating out of Tunisia. Palermo is on the northwestern coastline but was out of Allied fighter range. The Allies found the ports of Syracuse and Palermo the most attractive, and both met the needs of the respective armies in terms of ship berthing and other infrastructure.

Planning for Husky began on February 10 by a nucleus of planners, mostly from Allied Force Headquarters. The staff assumed the designation Task Force 141 (named after the number of the room used in the Hotel St. George.) The planning group was a subsection of AFHQ G3 but later would evolve to be the nucleus of the 15th Army Group staff.

The two principal Allied formations were Force 343 (Western Task Force, US Seventh Army) and Force 545 (Eastern Task Force, British Eighth Army). Planning for the operation was complicated—forces for the assault would launch from both ends of the Mediterranean as well as from the United States. Commanders and their planners found themselves spread across the theater, and there was still an ongoing battle in Tunisia, which tied up a number of units destined for Sicily.

Despite the efforts to learn from past mistakes, the planning for Husky had just as many problems as the planning for Torch, mainly because of the continuing fight in Tunisia and the inability to consolidate all of the different planning staffs into one general location. Hundreds of miles separated the staffs of the different organizations. For example, Force 141 (the future 15th Army Group) and the US Navy positioned their headquarters at Algiers. Force 343, I Armored Corps and later the Seventh Army, was at Rabat, French Morocco. The 3rd Infantry Division was at Bizerte. Distances were great, and travel was difficult— two factors that limited coordination and cooperation.[5]

The North African air forces found it difficult to dedicate much effort to the planning process because of ongoing combat operations against German and

Italian forces. Patton had little faith in the air corps and its ability to provide effective ground support. He even went so far as to ask Vice Admiral Kent Hewitt for air support from the navy: "You can get your Navy planes to do anything you want, but we can't get the Air Force to do a goddam thing." The invasion force sailed with no clear understanding of air routes for airborne and glider forces or whether the invasion force would have any support from the air forces.[6]

Senior commanders finally collectively focused on the plans for Husky at a commander's conference on March 18. The draft plan, dated March 3, called for split landings, with the British seizing the port of Syracuse and the Americans seizing the port of Palermo five days later. This plan would split Allied forces but would also provide each task force with a sizeable port facility, which could speed the buildup of the beachheads and facilitate the drive inland toward Messina—the campaign's objective. This course of action assumed that the poor and limited beaches of Sicily could not support a large invasion force and that each of the two armies needed a major port. As such, the driving factor for this initial plan was logistically oriented.[7]

Force 141 conducted the overall level of planning but left the details of the ground-assault plan up to the respective task force commanders. On April 7, Patton and Montgomery received a memorandum asking for their outline plans no later than May 1. Patton and his staff, at that time in French Morocco assigned to I Armored Corps, were ready to begin planning; however, Montgomery and the Eighth Army staff were still engaged in Tunisia and did not have the time or inclination to shape the early planning of Husky.[8]

British officers, including the head of the naval element of the Eastern Task Force, Admiral Bertram Ramsay, and the Eighth Army commander, General Montgomery, objected to the plan, believing that the two task forces needed to land in closer proximity to each other or risk isolation and destruction by Axis coastal units and divisions already on the island. General Alexander, the deputy commander in chief and commander of the 15th Army Group, called for conferences on April 27 and 29 to work through the differences.[9]

The conferences produced a revised plan on May 19, with a final outline plan dated May 21. The new plan dropped the seizure of Palermo and moved the American landings down to the shores of southern Sicily, immediately west of the British landing sites. The new plan sacrificed logistical supportability for the sake of a stronger beachhead. There was an element of risk in each option, so Eisenhower had to make the final decision. For each task force, the initial mission was essentially the same: to seize and secure beaches, ports, and airfields within its sector to enable the required buildup. Following that, each task force would drive north to secure the island.[10]

Torch had shown that conducting sustainment operations over the shore was typically much more difficult than planned and that established port facilities were essential to conducting a rapid buildup of forces ashore. The decision to move the American invasion site from Palermo to beaches west of the Eastern Task Force meant that the port of Palermo was no longer available and that sustainment planners had to look for other means to build up matériel to support the drive inland, which required a major modification of the Seventh Army's sustainment plans—all within seven weeks of the landings.

The loss of Palermo would be significant because many experts estimated that the port could support up to ten divisions. Now the Seventh Army had to sustain itself from across the beaches. Both Licata and Gela were in the American sector, and each had a small port, but the capabilities were limited and came nowhere near that of Palermo. Licata had an estimated daily capability of 600 tons per day, compared to Syracuse at 1,000 tons per day and Palermo at 2,500 tons. Scoglitti had only a small fishing port. Gela could handle a mere 200 tons per day. The loss of a large port put into question the theater's ability to resupply US forces.[11]

The creation of a single logistical headquarters for both the British and US forces was never a serious consideration. Due to the differences in systems noted during the Tunisian campaign, the distances between headquarters elements on North Africa and the embarked forces, and the separation of bases within Sicily, the British general Sir Harold Alexander decided to maintain the existing administrative support relationships. The US theater organization would support all forces operating in the American sector; the British would take care of their sector.[12]

AFHQ G4 and the Services of Supply began working on revised maintenance plans for the operation on May 23. The new support plans called for the three divisions within the Seventh Army to land with seven days of supplies combat loaded into assault craft. An additional fourteen days' worth of matériel would arrive in two follow-on convoys. The Cent Force (45th Infantry Division) was to land on the beaches near Scoglitti. The Dime Force (1st Infantry Division) was to land on the beaches surrounding Gela. Last, the Joss Force (3rd Infantry Division with a combat command from 2nd Armored Division) was to land on the beaches around Licata. Kool Force (the remainder of 2nd Armored Division) was a floating reserve, with twenty-one days' worth of supplies for its own use.[13]

This supply plan significantly increased the quantity of supplies needed to accompany the assault forces. Original plans had called for four to five days of supplies for the assault forces; now each carried seven days' worth. This meant that units had to modify requisitions as well as the respective load plans and landing plans. In a scene reminiscent to Torch, logisticians found themselves modifying support plans at the last minute to accommodate the combat forces' changing tactical plans.

Due to the lack of a tactical plan prior to debarkation, logisticians based their support plans on generic-type units in generic missions on generic beaches. This required some reloading prior to combat, so that many vehicles were not combat loaded. Only after the assault ships had sailed did Force 343 discover that its assault beaches had 1,000 yards of sand dunes behind the beaches, a terrain factor that complicated mobility and required engineer support to improve the beach exits. Had the task force known this before embarkation, the Seventh Army G4 would have planned for additional road and beach construction materials. The late identification of specific beaches stemmed from limited preassault reconnaissance because commanders did not want to flag their intensions.[14]

As in Torch, in Husky planning was not fully coordinated either between the combat and support commanders or between the services. Ultimately, the absence of tactical plans by the Seventh Army commander meant that logisticians had to develop a support plan that could support any tactical plan and that joint planning was next to impossible. The lack of a tactical plan affected more than just the theater. As one senior British administrative planner noted,

> I was under great pressure to furnish certain essential information regarding administrative requirements to the War Office without delay, and this meant that I had on more than one occasion, to produce the administrative plan for various phases before the operational plan had been prepared. . . . When our plans were eventually seen by the Commanders-in-Chief concerned[,] they were, as was of course natural, modified radically. This meant that a great deal of work had to be done over again; *but much more serious it upset forecasts that had already been sent to London and Washington in regard to the weapons and equipment that we had to have to make the operation a success.*[15]

Any changes to the administrative support for Husky could easily affect support to the other theaters of war as well, all of which were competing for limited resources.

Despite the changes, units finally solidified their plans. The Eastern BS was to serve as the reinforcing base section, with the main supply depots at Ferryville and Mateur, Tunisia. The base section headquarters operated out of Mateur. The theater SOS would load seven days of follow-on supplies into coaster vessels, which were planned to arrive in Sicily on D+14. For long-term support of the operation, follow-on convoys would carry supplies from the United States at fifteen-day intervals. The SOS also maintained an additional fifteen days of supplies in Tunisia as a reserve that could be moved on short notice. Any items needed by the Seventh Army that were not on the US convoys or needed before the next convoy would be sourced and shipped from North Africa by the theater. Service forces would accompany the assault force to land on the beaches and small ports in the US sector, but if these sites proved inadequate, US planners made provisions with the British force to share use of the Syracuse port beginning approximately D+14.[16]

During the assault phase of the operation, any requests for US resupply outside of the preplanned convoys were to flow from the units up to the Seventh Army. The Seventh Army Rear Headquarters, located at Bizerte, would review the requisitions and fill them from army stocks if possible. The army would pass any unfilled requisitions to the Eastern BS. From here, the base section would work to fill all possible requirements from stocks already in the theater. Critical items, such as repair parts for mechanized equipment, would fly to Sicily by plane, while other, less-critical items were to move by sea.

By May 1943, planning was in full progress, but an additional challenge faced the planners: units had never seen many of the different assault craft projected for use in Husky, so they did not have an appreciation of the capabilities and requirements of these new items. For example, there were no data available for craft such as the Landing Craft, Tank (Rocket) or the Landing Craft, Gun. No one had ever seen how many personnel could fit within a Landing Ship, Tank (LST). These pieces of information mattered a great deal because they had a direct impact on the planning of the loading and landing of the task forces. The exact size of assault-craft ramps dictated which types of equipment could fit onto a particular vessel. Personnel capacities dictated whether units could be transported whole or in pieces. Beaching depths determined whether infantrymen could wade ashore or needed inflatable boats.[17]

Back in North Africa, each base section commander was responsible for the support of Force 343 units staging in their respective areas, but once the units arrived in Sicily, the responsibility for support beyond the Seventh Army's capability would transition to the commander of the Eastern BS. To help facilitate coordination, Patton agreed to have a Force 343 liaison officer stationed with the Eastern BS—someone who could speak with Patton's authority for many questions dealing with personnel, equipment, or supplies for the US forces in Husky. General Larkin also stationed an SOS liaison officer with the Eastern BS to help coordinate any requirements that might be beyond the base section's capability.[18]

Whereas the SOS and the AFHQ G4 had a fair idea of the types and quantities of supplies needed by the ground forces for Husky, the Allied air forces never provided a specific list of needs. In general, there was widespread dissatisfaction with the lack of US Army Air Corps participation in the overall planning effort. There was no coordinated planning effort between the services, and the support plan provided by the air command was "the most masterful piece of uninformative prevarication, totally unrelated to the Naval and Military Joint Plan, which would possibly have been published." The air plan called for daily landings of massive quantities of supplies and equipment—but with no specifications regarding weight, size, or projected use. The air command apparently expected this level of support even though its resupply plan called for convoys every four days despite the fact that the theater was planning convoys for every fourteen days. AFHQ and the SOS conducted support planning in the cork forests and olive groves surrounding the Force 343 headquarters, but the Allied Air Forces declined to provide a representative.[19]

Ultimately, the lack of a ground tactical plan from the Seventh Army provided the biggest challenge for logistics planners. It meant that any logistics plan needed to be flexible enough to accommodate *any* combat plan. This same lack of a ground tactical plan also inhibited planning coordination among the joint services.[20]

Despite the lessons of Torch, Husky hardly served as a role model for joint-service planning. Perhaps this was understandable, given that operations were still going on in Tunisia, but it was not inevitable. The planning effort *did* improve in terms of anticipating the need for a more balanced ground force that included the necessary combat and combat-support elements. However, work remained to be done in integrating the air forces into the planning effort. Regardless, as summer drew near, the time for planning ended, and the time to load the invasion force came.

Mounting the Force

Brigadier General Arthur Pence received word in late April that the Eastern BS would have a large role in equipping and supplying Force 343 units for the invasion and for loading the invasion force and accompanying supplies on the assault craft, a process known as "mounting." At the same time, the base section was deeply involved in supporting the Allied drive across Tunisia. Short on personnel, the Eastern BS pulled two officers and one enlisted man from the staff to form a special planning group for the mounting of Husky.[21]

To complicate matters, the Allies did not capture the area of Tunisia needed for the mounting operation until May 10, 1943—less than two months before the convoy's departure date. Planners were unable to conduct reconnaissance trips or to survey potential support sites, so maps instead had to serve as the main sources of information. The Eastern BS was responsible for the US mounting effort. It was to equip all troops participating in Husky from North Africa, provide thirty days of supplies for these forces, transport units to the ports of embarkation, and then load them on assault vessels.[22]

The forces to be used for Husky were spread across French Morocco, Algeria, and Tunisia. The SOS had to gather these units at specified ports and transport them to the ports of embarkation following any training. Service units identified and packaged supplies for the assault and any follow-on support. Some cargo was preloaded on vessels, while other cargo was stored in Tunisia, awaiting later shipment. This undertaking was massive in terms of both complexity and magnitude and had to occur simultaneously with the normal administrative support still required in North Africa, including support of US ground and air troops, civilian populations, French forces, and POWs.

To support the force in Sicily and to respond to changes and unforeseen requirements, the Eastern BS developed an extremely flexible support plan. The general depot was at Mateur, Tunisia, due to its close proximately to air and ground bases, seaports, and the rail and road network. Truck units and rail moved supplies and equipment to either Bizerte or Tunis for sea transport. Although the British controlled Tunisia, AFHQ arranged for the Eastern BS to move its advance headquarters and a series of supply depots to Mateur.

The Husky mounting operation was a new challenge for AFHQ and the North African theater. The War Department or commanders of the European theater had prepared and mounted forces for the earlier Torch invasion. This was the first time the North African theater would mount a sizeable assault using its own forces, and there was virtually no information available on exactly how to conduct such an operation. The theater had learned much since Torch, but there were still moments of confusion. The past year had seen seven different directives on the labeling of cargo. The initial load plans for the Seventh Army had to be redone when the list of requirements for the US Army Air Forces arrived late. Units demanded priority and additional space. Considering all this, it is therefore surprising that the mounting operation occurred as well as it did.[23]

The two divisions that were already in North Africa, the 1st and 3rd Infantry Divisions, had little time for in-depth load planning based on the final version of the Husky plan. Logistics officers representing the various units and staffs gathered on the afternoon of May 20 to discuss the upcoming movement. Units had only forty-eight hours to develop their supply requirements. To meet the necessary timelines, supply requisitions for the assault force were due on the afternoon of Saturday, May 22. Requirements for the follow-on convoys were due two days later. The schedule obviously left little time for detailed planning, so supply planners used the experiences of the North African campaign to make their best judgments. Personnel from the 1st Infantry Division were to load transports at Algiers; the personnel for 3rd Infantry Division would load craft at Bizerte; and Tunis would serve as the main embarkation port for all Force 343 equipment.[24]

Major General Terry Allen, commander of the 1st Infantry Division, was worried that his units would not have sufficient time to fix their equipment before loading up for Sicily. Five months of hard fighting in North Africa had left the division's vehicles and weapons systems in a general state of disrepair.[25]

The Mediterranean BS responded, however, putting together a plan to overhaul the equipment of the 1st Infantry Division within ten days of receipt. The only deficiencies not repaired were those requiring repair parts not physically available in North Africa. The plan worked, and the division was able to meet its load dates with equipment that was in much better shape than expected. By the first week of June, the two divisions were now ready to move to the assembly areas and begin loading the assault craft.[26]

To conduct the actual mounting operation, the Eastern BS formed the 1st Embarkation Group (Provisional). No such group had existed previously, so it was very much a trial-and-error type of operation. The lack of experience and the constantly changing plans produced a difficult situation for the new unit. To provide additional labor for the mounting, personnel from the 34th (Red Bull) Infantry Division helped to establish and run the camps and vehicle parks.[27]

The advance party from the embarkation group arrived in the area on May 31. The assembly areas opened shortly thereafter, from June 10 to June 13. SOS and 34th Infantry Division personnel worked quickly to post signs, establish communications, stockpile food, arrange for water and sanitation facilities, establish medical facilities, provide security, and coordinate movement plans. The maximum troop capacity of the assembly areas totaled 25,000.[28]

The major landing craft included Landing Craft, Infantry (LCI), LST, and Landing Craft, Tank (LCT). The LCIs were seagoing craft that could carry between 180 and 200 soldiers at a time and land them directly onto a beach. The LSTs were the workhorses of the armada—seagoing vessels designed to carry large amounts of men, equipment, and supplies. Each LST could carry approximately twenty medium tanks and about 200 men. These vessels did not need a pier or port to discharge their cargo. The LST's smaller cousin was the LCT. Each of the flat-bottomed LCTs carried nine Sherman tanks or 300 tons of cargo. These vessels had only half the speed of the larger LSTs, however, and did not do well in lengthy open-sea crossings.

Small landing craft that could carry soldiers and also avoid obstacles formed much of the first wave of an amphibious assault. Landing Craft, Vehicles Personnel (LCVPs), also known as "Higgins Boats," were plywood craft that had flat bottoms and could carry up to 36 troops or a small vehicle such as a Jeep but were not made for long open-water voyages. Similar to the LCVP, the Landing Craft, Assault was a small plywood craft designed to carry 31 soldiers from the larger troop transport ships to the assault beaches. Together, these small craft landed many of the infantry units onto enemy beaches.

The embarkation group could simultaneously stage and load 52 LCIs, 30 LSTs, 32 LCTs (vehicles and equipment), and 12 personnel-only LCTs. Each task force moved into a staging area the day prior to moving into its assembly area. From there, vessel-size groupings of equipment, known as "serials," moved into the loading areas on the scheduled loading day. The first vehicle serials loaded on June 18; personnel serials started loading on June 23. The Cent Force, 45th Infantry Division, had loaded its transports earlier in the United States and sailed for North Africa, where it was to meet with the other divisions.

The final assault force to complete loading was the Joss Force of the 3rd Infantry Division. Loading for this force encountered problems in late June when the division's supplies arrived out of order at the assembly areas. The problem originated at the theater depots when service units failed to follow instructions and loaded the wrong supplies in trucks heading for the docks. To deal with the problem, the 1st Embarkation Group established dockside dumps, which allowed units to off-load, sort, and then load supplies on the correct vessels. This process slowed loading but not enough to hamper the timeline. The group completed the loading of supplies on July 5. Joss Force personnel were the last units to load, with units filling the LCIs on July 5 and the LSTs on July 6.

The mounting operation showed that the theater needed to allow at least sixty days between the completion of planning and the earliest sail date: nine days to complete load plans, thirty days to waterproof supplies and equipment, seven days to shift cargo to match the ship-specific transport quartermaster plans, and fourteen days to load a division.[29]

Coordination between the navy and army was still a problem. Army planners decided that mules would be useful in the hilly Sicilian terrain, but no one told the Joss naval commander of the plan, causing an altercation between Rear Admiral R. L. Connolly and Major General Lucian Truscott over a navy tradition of not carrying livestock. Following their spirited conversation, the navy agreed to carry anything the army needed.[30]

Waterproofing was still a developing skill. Each vehicle, radio, or other sensitive item of equipment landing with the assault force needed the right protection to keep seawater from getting into vital components, such as engines and batteries, during the voyage to the assault area as well as during the transition from the assault craft to the beach itself. Improper waterproofing often produced unserviceable equipment at the far shore. The US forces experienced problems because of such damage during Torch; now the Embarkation Group worked to provide a better level of protection for the Husky forces' equipment.

To better coordinate and synchronize these limited assets, the SOS and the Seventh Army modified their organizations. The Seventh Army added a large transportation section to the G4 in order to plan and direct supply activities of all army service forces. In addition to the 1st Embarkation Group (Provisional), Seventh Army formed a "near shore control group," a rear echelon, to ensure that the follow-on convoys from North Africa met the army's changing demands. Finally, the 1st Engineer Special Brigade would organize the beaches and serve essentially as an advance base section on the far shore of Sicily. These new organizations and their missions were the direct outcome of hard lessons learned during the Torch landings.

By July 5, the mounting of Husky was complete. With little experience or time for preparation, the Eastern BS had loaded 77,520 personnel, 17,165 vehicles, and more than 12,000 tons of supplies onto the Force 343 assault craft. For British XIII Corps alone, there were 4,300 vehicles, 191 tanks, and more than 11,700 tons of supplies. The Luftwaffe tried to interdict the fleet with an attack by seventy-five aircraft on Tunisian ports on July 6, but no ships were hit, and there was only minor port damage. The subsequent landings would determine whether this supply effort was sufficient and whether the organizational changes could overcome the problems experienced seven months earlier.[31]

Initial Assault

The assault of Sicily was the largest amphibious assault conducted up to that time, consisting of six US and five British divisions. The naval armada contained more than 3,200 vessels, 1,700 of which conducted the movement of the US force and its cargo. AFHQ envisioned three phases. The first phase was the seizure of the beachheads and airfields near the beaches so they could serve as bases for following operations. The second phase was the seizure and opening of the ports. The Seventh Army would secure the western part of the island, while the British Eighth Army would drive north to seize Syracuse, Augusta, and Catania. In the third phase, the Allies would complete the occupation of the island by using the British force to drive north along the eastern coast, while the Seventh Army moved east along the northern coast road. The campaign would be complete once Messina fell. With the Axis loss of Sicily, the Allies hoped to knock Italy out of the war.[32]

Unlike in Torch, the landings for Husky would be against Axis forces consisting of four Italian divisions and two German divisions, totaling some 300,000 plus troops. However, Sicilian defenders were hampered by a shortage of ground transport, a limited road network, mountainous terrain, and daily Allied attack by air. The Italians' will to fight was very questionable.[33]

The invasion started during the night on July 9, 1943, with US and British airborne drops planned for objectives behind the beaches. In the British sector, the Pointe Grande Bridge was the key to seizing Syracuse. Montgomery decided to use glider elements of the British 1st Airborne Division for this objective. Unfortunately, the landing zone contained stonewalls and large cliffs—obstacles that made the area wholly unsuited for the landing of the large, fragile glider craft.[34]

The 2,075 glider men set out from six airfields in Tunisia in 137 Waco and 10 Horsa gliders and headed east for Malta. Passing over the navigation aid of Malta, they assumed the final course for Sicily. However, a combination of poor night navigation by the tow pilots, improper distance estimation by the glider pilots, enemy flak, and 30- to 35-mile-an-hour winds off Sicily's coastline served to break up the formations. Sixty-nine of Operation Ladbroke's 147 gliders crashed into the Mediterranean. Fifty-nine landed over a 25-mile area. The rest returned to Tunisia or crashed due to friendly fire. Only twelve made it to the assigned landing zones. As a result, only eight officers and sixty-five men of the British Airborne "Red Devils" captured the Pointe Grande Bridge, holding it less than twelve hours before being overrun.[35]

The Americans' airborne operation experienced less loss of life but similarly failed to land many forces to achieve their objectives. High winds and low moonlight meant that crews had a hard time navigating to their assigned drop zones. Pilots lost their sense of direction and missed checkpoints because of haze, fires, and dust. Antiaircraft fire from Allied ships resulted in the loss of more than twenty aircraft and 300 paratroopers largely due to poor coordination. Aircraft flew and discharged their human cargo all over the southeastern part of the island. The 3,400 US paratroopers thus landed all over the southeast corner of Sicily, ranging from the Eighth Division's sector, west across Gela, to the 45th Division's beaches. Only one battalion, the 2nd Battalion of the 505th Parachute Infantry Regiment, landed relatively intact, but it also landed 25 miles from the expected drop zone.[36]

Despite the missed landings, US paratroopers made the most of the situation by disrupting the enemy's areas. They cut communication lines, destroyed supplies, ambushed patrols, and caused confusion among the Axis commanders regarding the locations of the main landings. Living off the supplies they carried or found, the paratroopers operated in small bands, doing whatever they could until they joined up with the divisions arriving over the sea. The beach assault began during the early morning hours of the following day, July 10.[37]

Compared to Torch, the Husky beach landings were a considerable improvement, although coordination between the navy and army was still a problem. Even though the Allies had to contend with a force spread across 100 miles of beaches as well as with strong winds and high seas, the assault occurred largely according to plan. Nevertheless, ships and landing craft sometimes landed at the wrong ports and beaches or without warning. Within the British sector, a lack of training among naval personnel resulted in some units landing between 1,000 and 6,000 yards off their intended target. As noted earlier, critical to the operation of the American beaches was a new type of unit: the Engineer Special Brigade.[38]

The 1st Engineer Special Brigade was to remedy many of the problems experienced in the beaches of North Africa. Redesignated from the original 1st Engineer Amphibian Brigade, the new brigade consisted of four shore groups (one for each US division) plus one smaller shore group designated for the floating reserve. The brigade had the mission of supporting all four divisions and supplying all Seventh Army units for a period of up to thirty days following the assault. Not only was the brigade to perform the engineering functions of opening up the beaches, but it was also to act as the de facto base section in Sicily until the SOS could flow sufficient troops in to assume the logistics mission.

Each shore group consisted of an engineer regiment plus other support units, such as a medical battalion, quartermaster battalion, naval beach battalion, as well as ordnance and signal companies. The SOS added other capabilities depending on the infrastructure in the assigned area, such as rail lines. The 36th Engineer Shore Group was the largest group in the brigade, totaling 4,744 officers and men.[39]

The impetus behind the creation of beach organizations was similar to that behind the creation of the theater base sections and the SOS in general—to take the burden of administrative issues off the ground tactical commander, which would leave the commander and his staff free to focus on the fight before them and to plan the next move. The Seventh Army G4 would still be responsible for logistic planning and policy, but until the SOS could establish a local base section, the 1st Engineer Shore Brigade had the responsibility to conduct the required sustainment operations.[40]

Each beach group had the responsibility to organize its respective beaches. This included the tasks of unloading ships, marking aids and obstacles to navigation, constructing beach exits, providing security, setting up beach dumps to receive supplies from the landing craft, unloading boats, and moving supplies inland. The groups also had to operate vehicle-maintenance areas to clear the beaches, run de-waterproofing areas to prepare vehicles and equipment for combat, and operate a casualty aid-and-evacuation station. Each shore group could operate up to six beaches and one divisional supply dump. A typical division required between four and six beaches, depending on the number of attached units. Generally, the navy had responsibility for the beach to the high-water mark. From there, the army and navy shared responsibility for the near beach (the strip of land close to the high-water mark that formed an area for beaching of vessels, ship-to-shore communications, anchorage, and initial

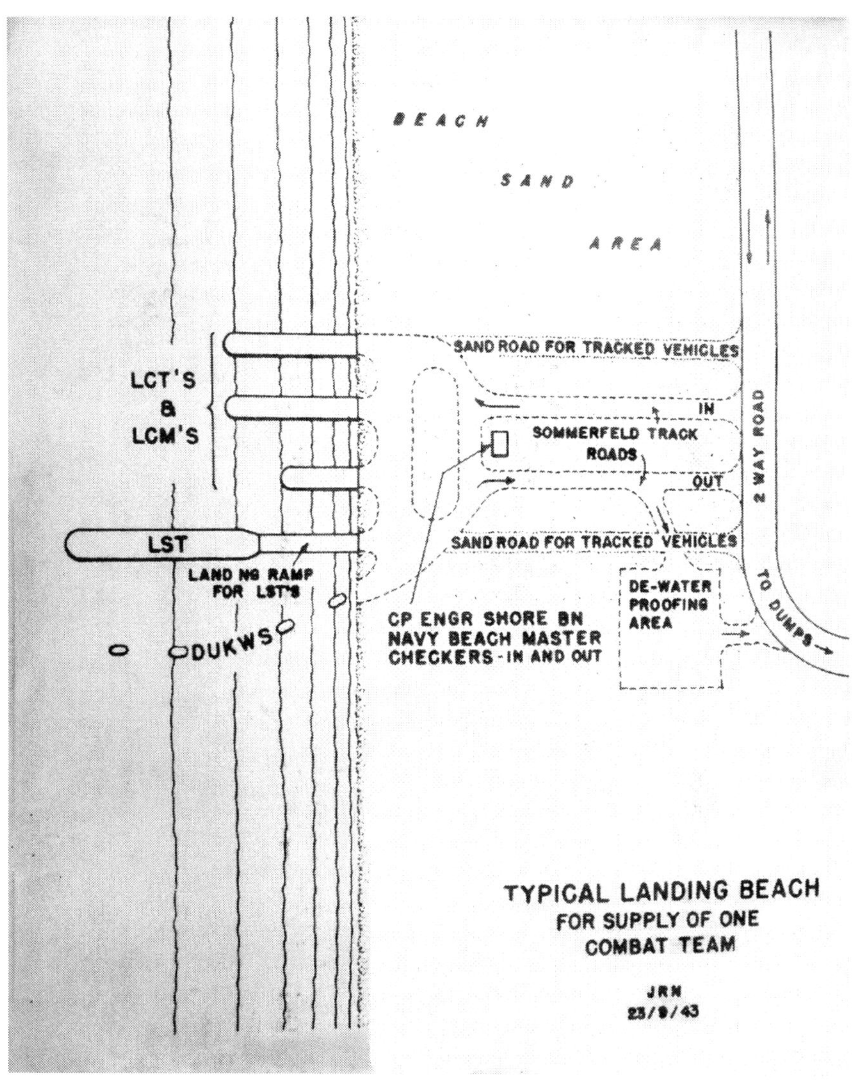

Figure 7. Typical landing beach for one combat team (National Archives, College Park, MD)

unloading). The army had sole responsibility for the far beach (the area inland of the shore that could be used for cover, clearance, and concentration of units).[41]

Although the Allies had gained insights from Torch, there were still surprises and challenges to overcome. Beaches south of Gela contained obstacles that threatened to impede sustainment of the force. Many of the beaches in the American sector included cliffs that limited or prevented vehicles from exiting

Figure 8. DUKWs at Scoglitti (US Army Quartermaster Museum, Fort Lee, VA)

the beach. The beaches themselves were about 30 yards in width, with sand dunes running in the back for 700 to 1,000 yards. The sand dunes, often made of soft sand or rocks, were large enough to stop vehicles.

At several beaches, such as Red Beach in the 3rd Infantry Division sector, a sand bar ran offshore parallel to the entire beach, thus limiting the approach of the larger assault ships. Beach exits to the island's interior were often limited or nonexistent. A few LSTs failed to unload all or some of their cargo—much of the signal equipment bound for Gela inadvertently returned to North Africa. Landings in the British sector were less than ideal as well. Transports were supposed to anchor 7 miles off shore, but anchored 12 miles instead, far outside the range of beach parties. Landing craft became lost in the darkness.[42]

Tanks and the new amphibious vehicles known as DUKWs ("ducks," a manufacturer's code rather than an acronym) were the only vehicles that had no difficulties in transitioning from the water to the beach. Although the new landing craft, such as LCTs and LSTs, had a flat bottom and could beach themselves in relatively shallow water, the combination of high ramps and soft sand caused many wheeled vehicles, especially those carrying heavy loads or trailers, to stall in the water or become stuck. In several cases, beaches became cluttered as LCVPs breached and became stranded and thus prevented the landing of larger vessels.[43]

As in Torch, there was still an absence of strong leadership on some of the beaches. The navy shore parties and the engineer shore groups were not well connected, and their actions were often uncoordinated. Traffic jams occurred in both the 3rd and 45th Division sectors due to the limited number of beach exits and limited road networks. Vehicle recovery in the surf and on the beach was slow and diverted other support vehicles, such as the DUKWs, from their assigned tasks.[44]

Beach units off-loaded much of the equipment and supplies by using an LST as a causeway onto the beach. Arriving LSTs anchored from half a mile to one mile off the beach. They would then transfer their cargo into the smaller LCTs, which could tie up at the offshore end of a beached LST. Roller conveyers on the beached causeway LST then enabled a quick loading of the LCTs.

The low beach gradients prevented the large LSTs from actually landing on many of Sicily's beaches. Knowing that this would be a problem, the navy devised a means to transfer vehicles from the LSTs onto the smaller LCTs, which could land on the beaches. While the LSTs were still in North Africa, the navy had cut access holes in their sides that were the same size as an LCT ramp. Once off Sicily, the LSTs anchored about 1,000 yards from the beach. Two LCTs then came alongside the LST. Workers lashed one LCT at a right angle to the side opening of the LST to serve as a loading platform. The second LCT then tied off to the first LCT. Drivers quickly drove vehicles off the LST, across the first LCT, finally ending up on the second LCT. As soon as this second craft was full, it untied and headed for the beach, with the empty space then filled by a waiting empty LCT. Working in this fashion, it took two and a half to three hours to off-load a full LST.[45] Ingenuity and cutting torches compensated for the lack of ports and fixed facilities.

Supplies were generally deposited on the beaches and then moved to inshore dumps, but in some cases essential supplies were taken off of LSTs, placed into DUKWs, and driven right to the front lines (during darkness), bypassing the dumps entirely. This method represented a vast improvement in support of the combat forces compared to the process used on the beaches of French Morocco and Algeria.[46]

Water was a critical commodity. A shortage of ammunition could stall an offensive, but a shortage of drinking water could bring Allied military operations to a rapid and total halt. Fuel kept vehicles moving, but in the infantry units water kept the combat forces moving, while ammunition kept them fighting. Water sources in Sicily were limited, and the daily temperatures in July averaged higher than 90 degrees Fahrenheit with little rain. Twenty water-tanker LSTs, with the accompanying storage points ashore, would help ensure that the Seventh Army could keep on the move.

One situation that could have easily impeded operations was the storage of food on the initial convoy. The assault wave carried twenty-four hours of rations in the easily unloadable LCTs, and the remainder of the food was stored on the rear of the LSTs. Poor weather conditions on July 8 threatened to affect vessel off-loading but calmed just prior to the landings. If the weather had not calmed when it did, sea conditions could easily have slowed the off-loading of the LSTs, which had the potential to quickly lead to a food shortage across the task force. Fortunately, the weather cooperated, and planners learned that in future landings the LCTs needed to carry more of a balance of supplies to allow for poor weather or critical supplies, such as food, belonged in the *front* of the LSTs to allow for easier access.[47]

The DUKWs were the recognized heroes of the Sicily landings for both US and British units. Used for the first time in actual combat, these vehicles proved themselves more capable than anyone had imagined. Due to the soft sand of the beaches, DUKWs ended up hauling the vast majority of supplies and equipment from the beaches to the inland dumps, two and a half tons at a time. They were hard to maneuver and slow, but they were unequalled in recovery operations and cross-beach mobility. Drivers could regulate the air pressure of the tires from inside the cab, allowing the vehicle to transition quickly between the beaches and the roads.[48]

Although the DUKWs excelled at transporting soldiers or cargo over short distances (up to a mile), the number of broken trucks quickly accumulated. Units, unfamiliar with the limitations of the equipment, diverted DUKWs from their assigned amphibious missions or used them to haul cargo over extended distances onshore. Manufacturers had not designed the trucks to cover long distances over roads or to travel across rough terrain. The diversion of these vehicles in the division areas quickly disrupted landing craft off-load plans and congested narrow roads along the beaches with derelict trucks.[49]

Because of the tremendous usefulness of these vehicles, maintenance issues also began to appear. The beaches were tough on tires and tubes. Propeller gear easily broke. And because these vehicles were relatively new to the army inventory, spare parts were still in short supply. Keeping these vehicles in full operation was an ongoing challenge for the beach groups. The original concept behind the amphibious vehicle was to use it as a reconnaissance vehicle for traversing rivers and lakes. At a demonstration on June 24, 1942, the commanding general of the Army Service Forces saw that they would be of invaluable service to logistics units and ordered the production of several hundred vehicles. A year later, the DUKWS became a central feature of the Husky support organization and enabled sustainment of the Seventh Army from over the beaches.[50]

Despite the inclusion of a greater number of vehicles with the invasion force, there were still not enough trucks to move all supplies and equipment forward.

Figure 9. The beach at Scoglitti (US Army Quartermaster Museum, Fort Lee, VA)

Fortunately, neither Allied air attack nor intentional destruction by Axis forces had damaged the Sicilian rail system to any great extent. By D+1, shore engineers and rail personnel had begun putting the rail system into use, pushing supplies forward.[51]

By the end of D+3, the Seventh Army had successfully landed 66,235 men, 17,766 tons of bulk cargo, and 7,416 vehicles on Sicilian beaches and the small ports of Licata and Empedale. The main beach supply dumps were located at Gela and Licata. In the 45th Division sector, beach units moved the original Green, Red, and Yellow beaches approximately three miles to the southeast, near Scoglitti, to take advantage of better beach conditions and an improved road network. Within ten days, the invasion was well under way. Most of the assault convoy had off-loaded, and the docks at Licata were in use. Engineers were working to improve the road networks. Food and water were an ongoing concern at the front lines due to limited transportation, but there was plenty of ammunition. By the end of July, the Seventh Army had landed 111,824 men, 104,734 tons of cargo, and 21,512 vehicles.[52]

Husky's beach operations were not perfect, but they had greatly improved compared to the North African landings eight months earlier. New equipment had solved many problems experienced during the earlier invasion, and commanders and their planners had done a better job in balancing the assault force. However, both the army and navy still had a great deal to learn. Both services needed firm leadership early on during the assault on *all* the beaches, not just on some of them. Another consideration is the fact that these invasions faced different conditions: the landings in Torch were largely unopposed or at the very least not strongly opposed, whereas the landings in Sicily were conducted in the face of a strong Axis force that was expecting a possible invasion.

In the grand scheme, the Allies were flooding ashore with all of the matériel needed to pursue a mobile, offensive campaign. This tremendous stream of supplies and equipment landing on Sicilian shores was not lost on the enemy. On July 11, an Italian homing pigeon accidentally landed on a US minesweeper. The message attached to it read in part, "Cargo ships by the hundreds are discharging uninterrupted war matériel. No friendly air. . . . Send more pigeons." The Seventh Army had successfully arrived on the beaches of Sicily and was ready to drive inland, heading toward the north.[53]

To the east, Montgomery and the British Eighth Army had faired a bit better on the landings. Whereas some of the US beaches experienced bottlenecks, mostly from a lack of beach personnel, the British were in a better position. Weather did not influence the beaches in the British sector as much as in the US sector, and Syracuse proved to be a very capable port. The British did have some difficulties in off-loading trucks as quickly as Montgomery wished, but, overall, the British support forces kept a steady stream of matériel moving forward.[54]

The presence of Syracuse in the British Eighth Army area aided Montgomery's sustainment operations. With a daily discharge capacity of 1,000 tons per day, the Eighth Army could send most of its matériel into Syracuse and use outlying beaches for any needed residual capacity. In addition, the port of Augusta, 15 miles north of Syracuse, would quickly add to British port capacities. None of the major unit combat reports indicates any big problems with the landing of the Eastern Task Force's 66,000 personnel, 10,000 vehicles, and 60,000 tons of supplies sent in on the initial assault. Eisenhower reported that by July 17 "Eighth Army's supply lines through the ports of Augusta and Syracuse were already in good working order."[55]

British beaches operated generally with the same type of support units as those found on US beaches. Shore parties handled the reception and organization of the beaches, although the British did have some issues with standardizing the makeup of the different beach groups, which included UK, North African, Middle Eastern, and Indian variants.[56]

The supporting British logistics headquarters was Tripbase, originally stationed out of Tripoli, but later renamed Fortbase (a part of the 15th Army Group) upon the transfer to Syracuse. The British Number 1 District provided logistics support for British forces on Sicily, while the North Africa District Headquarters maintained responsibility for support of the AFHQ British theater rear area in North Africa. The plan for two distinct systems of support was working.[57]

The Drive toward Palermo

On July 12, the Italian Sixth Army realized that counterattacks against the Allied beachheads were ineffective and that the time had come to switch to the counteroffensive. The commander of Axis troops on Sicily, General Alfredo Guzzoni, adopted a strategy to delay the Allied advance and begin withdrawing forces to the northeastern corner of the island. Patton, who was looking for an opportunity, became aware of the change in enemy intentions and quickly directed an aggressive move inland.

The II Corps commander, Major General Omar Bradley, moved two divisions toward the towns of Agrigento and Porto Empedocle on July 13. These towns were important strategically, and Porto Empedocle contained a port that could contribute to the support of the Seventh Army. Meanwhile, General Alexander was adjusting the US/British border to the west in an attempt to give Montgomery's Eighth Army an avenue of approach that could skirt past strong German defenses. This shift forced Patton and Bradley to adjust the US advance from the north to the west because there were only a few roads leading north, and Montgomery now controlled all of them.

Despite the change in directions, Patton still saw Palermo as the next big objective. Meeting with Alexander on July 17, the Seventh Army commander successfully argued that the best means to protect the Eighth Army was by aggressive action along the western portion of the island, thereby splitting the island in two. Alexander agreed, and the Seventh Army was free to resume its drive to the northwest.[58]

The southern towns of Agrigento and Porto Empedocle fell to the Seventh Army on July 16. Control of these towns gave the Seventh Army access to Highway 118, which led toward Palermo, and to a rail line that ran from Porto Empedocle north to Termini Imerse. These routes would become critical to sustaining the divisions as they moved north, away from the assault beaches. The Seventh Army's Provisional Corps, commanded by Major General Geoffrey Keyes, controlled the western part of the island, while Bradley's II Corps controlled the center.

The Provisional Corps had the mission of seizing Palermo, while II Corps moved north to cut the island in half. On July 19, the 82nd Airborne Division moved into western Sicily and began mopping up any enemy units that had not already moved east. To the east of the 82nd, the 3rd Infantry Division moved north along Highway 118 toward Palermo. The 2nd Armored Division followed as the army reserve. Resistance was light, and both divisions covered 20 to 25 miles. The combination of Allied air support and direct fire from the ground units was usually enough to convince Italian defenders to surrender without a fight. Patton's orders were to continue advancing toward Palermo unless ordered to stop. By the night of July 22, the 3rd Infantry and 2nd Armored Divisions were on the outskirts of Palermo, but the Italian defenders gave up without a fight.[59]

The fast capture of Palermo was due in part to the enemy's decision to shift focus to the northeastern corner of Sicily. Other factors included the bombing of Rome by Allied aircraft (thus redirecting the attention of the Italian High Command) as well as the sizeable stocks of fuel that were built up on the Seventh Army beaches. The drive was only about 100 miles in length, but three divisions on the move required large amounts of trucks and gasoline. The pursuit did not require much ammunition, but everything depended on mobility.

While II Corps was advancing northwest, the support situation was maturing as well. By July 17, the divisions began relinquishing control of their respective beaches and turning them over to the 1st Engineer Special Brigade, whose responsibility it was to gather all sustainment activities under its command. Within a week of landing, the brigade was now responsible for running the reception operations as well as for acting as the intermediate support element for the entire Seventh Army. The brigade assumed command over all shore groups and any service forces not assigned to the divisions or II Corps and began acting in the capacity of a base section for all US forces on the island.[60]

By July 23, Palermo was secure, and engineers began to repair the damage that retreating Germans had inflicted on the port. The damage was extensive but not prohibitive. The Germans had sunk thirty-four vessels and barges into the main shipping channels and along the docks. Working parties had to clear these obstructions before the port could be usable.

The port of Palermo was large enough to hold thirty-six LSTs and fourteen Liberty ships—a substantial improvement to the limited capability of the southern Sicilian beaches. Thanks to the engineers' focused efforts, the port was at 30 percent capacity within a week of its fall, but Bradley's corps meanwhile suffered significant supply shortages as it continued east toward Messina. The first Allied coasters arrived on July 28 carrying food, fuel, and ammunition. From Palermo, transportation units trucked supplies eastward to the advancing army. The capacity at Palermo continued to improve as the campaign progressed, providing a valuable means of supplying the Seventh Army's drive east toward Messina across the northern coast road. Patton had captured Palermo and now had access to a port along the northern Sicilian coastline, but this was only an intermediate objective. The real prize was Messina.[61]

The Race for Messina

On July 25, General Alexander formally assigned the Seventh Army the mission of helping the Eighth Army seize the city of Messina. Up to this point, this had been an Eighth Army objective but one that Montgomery's forces had difficulty achieving because of a strong German defense and the problems associated with going around the imposing Mount Etna. The capture of Palermo allowed Alexander to split Sicily into two sections along the east–west Highway 120. The Seventh Army had the northern portion, the Eighth Army had the southern.

The opportunity to grab Messina before the British could do so provided a powerful motivation for the American Seventh Army commander. Patton placed a renewed urgency in moving east along the northern coastline as quickly as possible. This urgency stressed not only the battalions, regiments, and divisions but also the supporting service elements as well. German engineers were adept at demolishing bridges and roads as they slowly retreated east, delaying the Allied forces' advance until Seventh Army engineers were able to repair the damage or construct new routes. Whether Patton's push for Messina represented personal ambition or merely the work of a demanding tactical commander is a topic for debate. Despite the increased casualties and risk, however, the Seventh Army steadily advanced.

Bradley ordered II Corps to advance toward Messina along a two-axis advance. Highway 113 skirted the northern coastline and provided a direct link with Palermo. Highway 120 ran about 20 miles to the south, linking Nicosia, Troina, and Randazzo. Between the two highways was a system of mountains and streams that effectively separated the American forces and prevented one unit from reinforcing or resupplying the other. The terrain favored the enemy and amplified any supply problems, which effectively slowed down the Seventh Army's movement eastward.[62]

For the supporting service units, the terrain not only tended to limit movement to the roads and rail networks but could also add to the time required to complete a convoy. Hills, defiles, valleys, and ridges served to slow down the rate of march. On one occasion, a II Corps convoy needed twenty-six hours to cover a 60-mile distance—an average of less than two and a half miles per hour.[63]

As divisions moved farther to the east, away from the beaches and Palermo, the line of communication lengthened as well. Supplies from Palermo moved east via rail, trucks, and watercraft to Termini, San Stefano, Brolo, and Barcellona-Calderà. The Seventh Army had difficulty pushing sufficient supplies forward to meet all the demand, so in some cases the army's divisions had to send trucks back to the supply dumps to draw supplies, a distance of up to 120 miles. Following doctrine of the period, the army and divisions had assigned support forces, but the corps did not, which saved the corps from expending energy and time on administrative matters but also meant that divisions had to skip an echelon if they needed to go to the rear for support.[64]

The steady resupply of ammunition to the forward US battalions and regiments proved to be a deciding factor in many engagements. The advance slowed when US units faced barrages from German artillery and supporting mortar fire. Working to seize Pizzo Spina, the 180th Infantry stopped the German artillery attack with a 100-round, thirty-minute counterattack of high-explosive and white-phosphorous rounds. In total, the 171st Field Artillery Battalion fired 1,100 rounds in support of the 180th Infantry on July 25, while a sister artillery battalion, the 189th, fired an additional 500. Such intensive attacks consumed large amounts of munitions, but they held the enemy at bay and ultimately decreased the number of friendly casualties. Logistics weighted the odds of a positive outcome in any battle.[65]

Because of the limited road and rail networks stretching across the island, the Engineer Special Brigade initially supplied the entire II Corps from just one army-level supply dump at Licata until the capture of Palermo. After that, as II Corps turned east toward Messina, service units established other supply points along the northern coast road, including at Caltanissetta, Petralia, and Nicosia and on the coast road itself.[66]

By D+20, sufficient army-level service forces were ashore to begin relieving the 1st Engineer Special Brigade from its base support mission. The brigade had performed admirably in executing its traditional engineering mission as well as in coordinating support. The Seventh Army began taking responsibility for the advance supply points and dumps, with the service personnel transferring from engineer control to army control.[67]

The key to the German defense was Major General Eberhard Rodt's 15th Panzer Grenadier Division. The 15th succeeded in slowing the US advance east on Highway 120 by defending all high ground and counterattacking whenever possible. On July 30, the 1st Infantry Division had reached the town of Troina, located along Highway 120 and the highest town in Sicily. Terrain here was difficult: hills surrounding the town held defensive positions. Mines littered the streams. Troina was important to the Germans because it was the northern anchor of the German Etna defensive line.[68]

Major General Terry Allen expected to take Troina without much of a fight, but US intelligence had failed to accurately assess the strength and determination of the German defense. Starting the attack on July 31, the 39th Infantry Regiment made some initial gains, but the 15th Panzer Grenadier Division, amid heavy enemy artillery fire, repulsed the regiment. Allen ordered the 39th and 26th Regiments to attack once again on August 1, but the US advance gained less than a half mile. Because of poor roads and heavy traffic, the supporting US artillery did not arrive in time to support the attack.[69]

Colonel George Taylor's 16th Infantry Regiment made another attack on August 3, but with only limited success. A counterattack in the midafternoon threatened the advance, but US artillery support prevented the Germans from overrunning the lead US battalions. The main US attack occurred on August 4. Determined to take the town, Allen pounded Troina with eight artillery battalions. The air force contributed as well, using seventy-two A-36 Apache fighters to bomb the German defensive lines. The 60th Regiment from the 9th Infantry Division arrived as well—providing fresh troops to augment the now-tired regiments of the 1st Infantry Division. By the end of the day, Allied air attacks had destroyed most of the German supply dumps. The Germans evacuated Troina on August 6, leaving 1,600 dead behind. The US force was now threatening to cut off the only escape route leading to Messina.[70]

The Axis evacuation of Sicily was one of the great logistical feats of the war. Faced by two Allied armies and their considerable resources, the German objective turned from an effort to push the Allies off the island to an evacuation operation—with the goal of safely transporting as many men and as much equipment as possible back to Italy. The senior German officer on Sicily, General Hans-Valentin Hube, ordered the evacuation of Sicily on August 4, 1943.

Responsibility for managing the retrograde across the Strait of Messina fell to Fregattenkapitaen Gustav von Liebenstein, a naval reserve officer and son of an army major general. Liebenstein quickly increased the daily capacity of the flotillas tenfold and increased the number of ferry routes. The priorities were men first, equipment and supplies second.[71]

Despite a half-hearted Allied attempt to interdict the strait, the evacuations occurred as planned. By August 16, the Italians had evacuated between 70,000 and 75,000 men, 500 vehicles, and 12 mules. The Germans evacuated 39,951 soldiers, 14,772 casualties, 9,789 vehicles, 51 tanks, 163 guns, and 18,665 tons of supplies and equipment. Only one German soldier died in the process. The evacuation was a masterwork of coordination, skill, and transportation management. The Axis withdrawal was aided by a reluctance on the part of the Allied air forces to employ heavy bombers against the Straits of Messina, resulting in a hollow victory. Sicily was now in the hands of the Allies, but most of the Axis defenders were safely ashore on the Italian mainland. A British War Office after-action report offered an appropriate summary of Husky as "a chaotic and deplorable example of everything that planning should not be." Despite the operational outcome, however, the sustainment effort was much improved from that of Torch.[72]

Supporting the Fight

Native labor came in many shapes, and it was not always in human form. The Allies had discovered the benefits of pack animals in the mountainous areas of North Africa, and they carried this lesson over to Sicily. As the Seventh Army moved across the difficult terrain of the northern coast, the need for such a capability became even more evident. Damaged or nonexistent roads meant that service units had to transport supplies either by hand or by mule. Teams of acquisition officers purchased mules at various locations and by various means. Some came from Sicily; others came from North African markets and were forwarded to Sicily by sea.

Eventually, more than 4,000 horses, mules, and donkeys were in use across Sicily. The 3rd Infantry Division alone used 650. These animals helped haul supplies, such as food, ammunition, and water to the dismounted combat units fighting in the rough terrain and then returned to the rear lines carrying the dead and injured.[73]

Besides trucks and pack animals, the use of existing rail lines also provided a valuable means of moving men and matériel across the island. Rail support began early after the landings. In sharp contrast to the Torch landings, in the Husky landings rail personnel from the 727th Railway Operating Battalion landed with the assault force and immediately went to work locating the rail yards and organizing local Sicilian workers. Within four hours of landing, the battalion was conducting a test run of the rail system around Licata. By D+1, supplies were being carried by rail to the 3rd Infantry Division, carrying 40 tons on the first day. Within three days, the daily tonnage had increased to 800. The mining and sabotage of the rail system that had been a problem in North Africa repeated itself in Sicily. Using techniques carried over from Tunisia, the 727th pushed an empty flat rail car ahead of the locomotive to find any hidden explosives on the rail lines. Rail workers had to clear spurs by hand. As the combat forces drove north and then east, the rail battalion followed in close pursuit.[74]

Initial rail service started out of Licata with 30 rail cars per day. Within a week, US rail service included 16 locomotives and more than 100 rail cars, now operating from Caltanissetta. Eventually, Sicily was able to provide more than 300 locomotives and 3,500 rail cars. The rail battalion established service up the center of the island and then spread it across the northern coast as the army turned east.[75]

Perhaps the greatest logistics challenge of Sicily was medical logistics. Ambulances were in short supply. There were only 3,300 hospital beds available for a force exceeding 200,000, thus requiring the evacuation of minor cases to North Africa for treatment. There were shortages of routine items such as blankets and splints. The Sicilian heat led to dehydration, heat injuries, and intestinal problems, sapping units of combat strength. Malaria began to take hold as well, infecting 10,000 soldiers in the US Seventh Army and almost 12,000 in the British Eighth Army. Despite these challenges, the campaign marched along.[76]

By September, the fighting had ceased, and US forces had withdrawn to the eastern and center parts of the island. As soon as rail lines and other infrastructure were no longer needed for military use, the Island BS turned these facilities over to the military government and local Sicilian authorities for continued maintenance. The 727th continued to coordinate all US rail support until it departed Sicily on October 27.[77]

The Island Base Section

Recognizing that Sicily would likely become a base of operations in the western Mediterranean, AFHQ decided to form a new base section on the island to provide intermediate support. Accordingly, the theater activated the 6625th Base Area Group (Provisional) on July 17, 1943. As part of the initial assault, the 6625th attached thirty-nine officers and ninety-eight enlisted personnel to the 1st Engineer Special Brigade to serve as part of the brigade during the landings and to act as the base section advance element. This advance element landed in Sicily between D+2 and D+9 and began work as part of the Engineer Special Brigade's staff.[78]

The 6625th was operating as a coherent support unit by August 18, although consolidation with the element at Licata would take a bit longer to sort through. With the formal arrival of the base unit in Sicily, the Seventh Army turned over all supply functions to the 6625th and the SOS, a responsibility Patton was glad to relinquish. The Seventh Army could now focus on combat operations, while the 6625th handled the administrative matters. Equally glad was the 1st Engineer Special Brigade. The establishment of the 6625th relieved the engineers of the supply and transportation mission effective August 25, allowing them to focus on engineer-specific tasks and missions.[79]

On September 1, the 6625th provisional Base Area Group became a formal base section, the Island BS, under the theater SOS and commanded by Colonel Robert Sears. Initially, the Island BS had responsibility for the western part of the island, while the Seventh Army maintained control of the east. Eventually, the Seventh Army transferred responsibility for the entire US sector of the island over to the Island BS. By November 1943, the Island BS controlled a swath along the northern Sicilian coast extending east from Palermo to Termini as well as the depots at Agrigento, Porto Empedocle, and Licata. The British Number 1 District controlled the rest of the island.[80]

Although the Island BS was established relatively late in the operation, the creation of such an organization made tactical sense and followed established doctrine. The Seventh Army was ready to shed itself of administrative oversight for the force, and by the middle of August there were adequate service troops on the island to accept the mission. Perhaps even more important, with the activation of the Island BS, the Seventh Army was now free to focus on any follow-on missions. The combat phase of Husky ended on August 17; Patton and his staff were looking ahead. The SOS and its Island BS took on the unglorified task of reequipping the divisions, mounting forces for the upcoming Italian invasion, establishing Italian service units, and closing out support facilities on Sicily. The 3rd and 45th Infantry Divisions had to prepare for the assault on Italy. The 1st and 9th Infantry Divisions and the 2nd Armored Division were destined for the United Kingdom to prepare for the upcoming cross-channel assault. Sicily had become part of the theater communications zone.

Assessing the Engineer Special Brigade Concept

Although the concept behind the Engineer Special Brigade and its beach groups seemed sound, Husky illustrated the challenges of implementing this new idea. The engineers and their attached units represented a unique capability, one that several divisions found difficult to leave on the beaches and shore areas and so instead put them to use elsewhere. Reports intentionally leave out the specific identities of the culprits, apparently to protect the guilty (and because division commanders were general officers). However, of the three divisions that landed at Sicily on D-Day, two used the engineer shore groups inappropriately.

One division chose to strip the service forces out of its beach group and to establish a separate division supply service farther inland. This ad hoc organization was not effective—it did not fit into the overall logistics plan, and the removal of the service forces severely hampered the operations of the beach group and its shore regiment. A second division indiscriminately used the shore group troops and vehicles to move supplies as far forward as the front line, thus impeding beach operations. Personnel of every rank were on the beaches commandeering vehicles and drivers. Divisional staff officers diverted and kidnapped entire convoys. Officers and men scrounged the beaches looking for specific supplies and equipment. The 1st Engineer Special Brigade had helped fix the assault beaches, but until the divisions learned to leave the shore groups alone to do their jobs, the beaches would continue to have problems.[81]

The navy had its complaints about the shore groups as well, claiming that beach conditions were chaotic from the time of the initial landing until D+3. The beach parties were overwhelmed by the number of boats arriving so quickly, and many vehicles ended up disabled on unmarked minefields. Beach parties failed to organize supply dumps, and so supplies piled up on shores. At night, the navy could not find shore-party personnel. Overall, the navy characterized the landings in Husky to be as disorganized as the landings during Torch the previous November, especially on the Cent and Dime beaches. Clearly, work remained in organizing future landings.[82]

Patton, following the conclusion of Husky, showed little change in his attitudes concerning logistics and administrative support. In October, he noted in his report that "in any landing operation confusion is bound to occur. . . . [L]anding craft do not land where they are supposed to and stores get ashore regardless of progress. The main thing is to get the men and stores ashore. If they are there, they can be used. If we waste valuable time trying to get them ashore in some preconceived order they will not be available."[83]

At least part of the problem, in Patton's view, was the type of officers commanding the various beach parties. Patton recognized that these parties could be of immense value in bringing order to the landing beaches; however, the beach parties "very seldom produce the results they should. This is due to the lack of force of character in the men of the Army and Navy commanding them." The services needed to get leaders on the beach who either through strong will or sufficient rank or a combination of both could provide direction and make things happen.[84]

Patton had his faults, but he was a good judge of character, particularly in issues involving questions of leadership. The Seventh Army commander was exactly right—the services were not staffing the beach parties with the best and brightest officers available, so the beaches were not as organized and beneficial as they otherwise might be. The beaches needed strong and forceful officers to bring order out of the chaos that was inevitable when trying to land hundreds of thousands of men and their associated equipment onto foreign soil in the face of the enemy. The beach commanders needed to control what was coming onto the beaches, direct where it landed, and then ensure it made its way to the right location past the beach. At Sicily, the Seventh Army expected the Engineer Special Brigades to control the beaches *and* support the divisions. This massive task demanded exceptional leadership and management skills. Unfortunately, most of the officers with those traits found themselves assigned to units other than the beach parties. Glory was still at the front, not in running a beach in the rear.

The Seventh Army commander was worried about only one thing: getting everything ashore as quickly as possible and then working it out from there. This concept works well for smaller amphibious landings, but for large landings involving myriad different units and types of equipment it is a recipe for confusion and inefficiency. Fortunately, Patton seemed content to leave his logisticians alone unless there was a problem that affected the army's operations. It is not that Patton did not care about logistics but rather that he preferred to focus on tactics and leave the administrative details for others to deal with.

Throughout July and early August, the Seventh Army found itself in a state of continual logistical evolution. The reception of matériel was transitioning from the beaches of southern Sicily to the port of Palermo. Responsibilities had shifted from the divisions to the 1st Engineer Special Brigade and the army. However, Sicily was not the only area undergoing change—the service forces in North Africa were transforming as well to meet the demands of the current fight and to prepare for the next one.

Adjustments

While AFHQ focused on the progress of operations for Husky, the theater SOS found itself confronted with a complex situation. The base sections had to support Sicily for any item not loaded on the preplanned convoys from the States. They also had to continue normal support for the hundreds of thousands of men still in North Africa as well as to prepare for the mounting of Operation Avalanche, the invasion of Italy. Given the lead times required to submit requisitions to the United States, planners had to work months in advance of anticipated needs.

The mounting of the Husky force had put the North African base sections out of balance. The Mediterranean BS and Eastern BS had issued large quantities of supplies and equipment to the divisions heading off to Sicily. By the end of June, these two base sections had to deal with shortages, while the Atlantic BS had excess stocks. In addition, the overall reduction in the number of troops in North Africa required the SOS and AFHQ to reassess the quantities of supplies needed within North Africa and to determine what might be available for support of Husky. The communications zone needed to adjust to the new environment.[85]

To accommodate the changing situation, the SOS decided to consolidate the majority of the theater's intermediate supply activities within the Mediterranean BS. Transportation officers diverted to the Mediterranean BS cargo that was originally destined for the other base sections. The Atlantic BS and Eastern BS would slowly draw down their stocks while shipping excess to other locations. Logisticians estimated that it would take five months to redistribute all excess stocks from the Atlantic BS.[86]

The complexity of supporting a military theater increased as the number of different items of equipment used by units increased. By 1943, more than 350,000 different items were in use across the theater, with more than 2,100 new items received in the month of June alone. All of these items needed close management to ensure that they ended up in the right units. In addition, new equipment required new repair parts and repair skills, thus increasing the complexity of tasks for quartermaster and ordnance units. The War Department was working to ship new items to the theater as quickly as possible, but the theater had to understand how to distribute, use, and maintain all this new equipment.[87]

Despite the effort to ship trucks to the Mediterranean, there continued to be a shortage of transportation in the communications zone throughout 1943. To deal with the constraints, the SOS became proficient at temporarily consolidating truck units under the G4 transportation section for centralized management of the fleet. In one such operation, the Eastern BS used eleven officers and 440 enlisted men to move ammunition over a thirteen-day period, operating twenty-four hours a day. Experiences of this type occurred frequently in the lead-up to Husky and helped shape later surges of transportation across the Mediterranean and European theaters. Later vehicle surges, such as the Red Ball Express in northwestern Europe in 1944, were rooted in the lessons and experiences of North Africa and Sicily. What made these types of actions possible was the existence of a centralized administrative headquarters that could assemble and manage the limited resources available to the theater commander. The North African theater was showing that it could, in fact, learn and adapt, and not all of the Allied victories were on the battlefields.[88]

Axis forces had few sustainment issues in Sicily because the island had already been serving as a supply base, and the distances to the support areas in Italy were relatively short compared to the distance to North Africa. However, Generalmajor Max Ulrich described a situation that impeded German commanders not only in Sicily but in all theaters involving German troops—the great number of different types of vehicles in German formations. Within the 29th Panzer Grenadier Division, there were almost 400 different types of vehicles. This diversity had implications for maintenance and spare parts, which constituted a "severe handicap to smooth marches." As a result, retreating German convoys had to stop every two hours to refuel and address maintenance issues.[89]

The fall of North Africa and the Allied invasion of Sicily did nothing to enhance Mussolini's reputation with the people of Italy, King Emmanuel, or the politicians in Rome. To make matters worse, on July 19, 1943, more than 500 Allied bombers struck Rome on the same day Mussolini met Hitler at Feltre. The attack heavily damaged a working-class neighborhood and inflicted heavy damage on San Lorenzo, one of Rome's seven basilicas. Following this attack, 150,000 Romans reportedly left the city. The people of Italy now placed their fate more in the hands of the church than with the government.[90]

Additional attacks the next week on the Italian mainland at Bologna and Foggia further undermined Italian support for the war and for fascism. Two days after the fall of Palermo, the Fascist Grand Council of Italy called Mussolini to stand before it. After twelve hours of debate, the council voted nineteen to seven in favor of no confidence in il Duce. General Vittorio Ambrosio, the head of Comando Supremo, Field Marshal Pietro Badoglio, a Mussolini opponent, and King Victor Emmanuel III combined forces to rid Italy of Mussolini and the Fascist government.[91]

Mussolini's removal from power generated a great deal of discussion and speculation among Allied leaders. The Americans argued that a stronger strategic bombing campaign of the Italian mainland might be sufficient to push Italy out of the war without an actual invasion. However, the British felt that only an actual invasion would convince the Italians to surrender. The Combined Chiefs of Staff held a meeting on July 26, during which they agreed to authorize Eisenhower to launch an invasion of Italy, Operation Avalanche, at the soonest possible date. If the threat of an invasion did not convince Italy to surrender, then perhaps the actual landings would.[92]

Back in Germany, the sudden removal of Mussolini surprised Hitler. At the Feltre Conference, the Italians had agreed to commit four additional divisions in southern Italy. By the following day, however, the situation was clear—Italy had revolted, and Mussolini was gone. Hitler quickly recalled Rommel and gave him command of Army Group B, with the mission to defend northern Italy. Meanwhile, the German 305th and 44th Infantry Divisions moved toward the Italian border and secured the Brenner and Mount Cenis Passes.[93] Germany prepared to take over its reluctant partner. Throughout the rest of July, the Germans worked to stage divisions in anticipation of an Italian double-cross, while the Allied leaders tried to find a diplomatic means of convincing Italy to surrender. Meanwhile, an additional impact of Husky was the opening of the Mediterranean to Allied shipping, which allowed the Americans to more safely send aid to the Soviet Union, thus "opening the floodgates of Lend-Lease."[94]

At the end of July 1943, two German divisions were already in Southern Italy, and 200,000 more Germans were assembled near Innsbruck. The Italian government felt itself to be a prisoner of the Germans and suggested that the Allies land in the Balkans and northern Italy as soon as possible. The fight for Sicily ended two weeks later, but the fight for Italy was just beginning. The Allies would have more opportunities to put the lessons of recent campaigns to the test.

5

Operation Avalanche

Italy

> How is it that the plans of two great empires like Britain and the United States should be so much hamstrung and limited by a hundred or two of these particular vessels will never be understood by history.
> —Winston Churchill to General George C. Marshall on the shortage
>   of LSTs, April 16, 1944

The Allied planning effort for the invasion of Italy had been in progress since the end of July 1943. Eisenhower focused on an attack around the Naples area; however, in late August and early September the exact position of the Italian government was still unclear. King Emmanuel wanted to remove Italy from the war but feared German reprisal. The staging of Rommel's Army Group B, almost 200,000 German forces, on the northern Italian border with Austria did nothing to calm the king's concerns, and the Allies were still in Sicily and North Africa.

The Allies remained divided in their strategic goals. Churchill wanted to focus on the Mediterranean approach into Europe, while Marshall feared additional efforts could divert resources away from a cross-channel assault. At the Quebec Conference in August 1943, Eisenhower received approval to invade Italy but with fewer forces than AFHQ desired.[1]

The tactical situation prevented a direct Allied movement on Rome. German forces had possession of Rome's airfields, preventing an Allied airborne operation. The Mediterranean coastline near Rome was beyond the range of Allied bombers but within range of Axis aircraft. The Italian government had held off a public announcement of the armistice terms, fearing a German invasion. Given these factors, Eisenhower decided to go ahead with landings at Salerno because they had the best chance of success.[2]

By September 6, Vice Admiral Kent Hewitt's Western Naval Task Force containing the Salerno invasion force was steaming east, toward Italy's western shore, uncertain of what it might find. The Italian government had privately surrendered three days earlier but publicly remained at war. Allied commanders hoped for a peaceful landing; they wanted Italian support, not resistance at the landing sites. The convoy came around the west coast of Sicily and picked up the

British 46th Division, already loaded into landing craft. The force continued its way east toward the Gulf of Salerno in calm seas.

At 6:30 p.m. on September 8, the ships tuned their radios to an Algiers station and turned on the ship's loudspeaker systems as General Eisenhower announced Italy's surrender to the world, which forced Marshal Badoglio to confirm publicly the armistice terms an hour later in Rome. The world now knew of Italy's capitulation, and the Germans were quick to react, sending Rommel's Army Group B across the Alps to reinforce the North of Italy just as the Allies were landing in Italy's South. Both moves set the stage for the battle of Italy.

Planning the Invasion

Planning for Operation Avalanche originated as a plan to attack Sardinia. As Patton and the Seventh Army were preparing for an attack on Sicily, Eisenhower ordered Lieutenant General Mark Clark, commander of US Fifth Army, to start planning for an assault on Sardinia, with the possibility of shifting the focus to the Italian mainland. Clark, age forty-seven, was the youngest lieutenant general in the US Army—a fact that annoyed Patton, who was eleven years older but achieved three-star rank after Clark did.[3]

The Fifth Army planned for a variety of possible operations. Options included Operations Brimstone, the attack on Sardinia; Musket, the attack on Taranto; Gangway, the attack on Naples; Barracuda, the attack south of Naples; and Avalanche, the attack on Salerno. The various efforts acquainted the planning staffs with the different areas of southern Italy, but the diverse plans also diluted the planning of what became the one final option—Avalanche. Eisenhower did not settle on the final landing location until July 27.

After much discussion between the Allied partners, the Salerno–Naples–Foggia area was chosen as the initial objective as part of Operation Avalanche. Salerno was just inside the range of friendly fighter aircraft (staging out of Sicily), and the port of Naples had sufficient capability to support a large army. Foggia was perfect for the basing of aircraft.

Eisenhower's plan was to establish a foothold on the Italian Peninsula, work north toward Rome, and from there continue north into the Po River Valley. This would draw German divisions away from the Western Front. The plan called for three different attacks. The British Eighth Army would mount from Sicily, cross the Strait of Messina, and seize Reggio di Calabria on the toe of Italy. In addition, a British airborne division would seize the port of Taranto on Italy's heel. Afterward, the Fifth Army, under the command of Lieutenant General Clark, would sail from North African ports and land in the Gulf of Salerno along Italy's western coast. Commanders judged a landing directly at Naples to be overly risky due to the level of German defenses there.[4]

The terrain surrounding the Gulf of Salerno presented challenges for the invading forces: there was only a small coastal plain, bordered by the sea on the west and by moderately high mountains from a distance of two to ten miles inland. Two narrow gorges would serve as chokepoints on the way to Naples. The mountains and rivers along the coast favored the defense and limited avenues of advance. Two major roads led north toward Rome. However, the beaches around Salerno were excellent for an amphibious assault, permitting landing craft to come close to the shore for discharge of men and matériel.[5]

The terrain leading north along the Italian coastline was similar in nature. Hilly and mountainous terrain in front of the northern Apennines provided excellent positions for German defenders. Ridge formations paralleled the coastal plain. Steep slopes and outcroppings limited mobility, which favored infantry rather than large tank formations.

On July 27, 1943, AFHQ issued the Fifth Army a directive to begin the planning for the attack on Naples. The draft plan was due to AFHQ by August 7, with an expected invasion date of September 7. Clark's plan called for the landing of two army corps, totaling 125,000 men. After landing at Salerno, the force would move north to capture the port and airfields surrounding Naples. Clark expected VI Corps to capture the port by D+12 but made allowances to operate the beaches for up to thirty days if necessary. Within twenty-five days, the Fifth Army would grow to 225,085 men and would land 45,262 vehicles and 153,930 tons of cargo.[6]

The Fifth Army was to contain two corps. VI Corps initially held the 34th Infantry Division, with elements of the 3rd, 45th, and 34th Infantry Divisions as well as the 82nd Airborne as follow-up forces. The addition of British X Corps, with its 46th and 56th Divisions, made the Fifth Army a combined Allied force. French forces would eventually land later in December, further complicating the logistical picture. For Operation Baytown, the invasion across the Strait of Messina, the British Eighth Army had the British 13th Corps, with two divisions: the 1st Canadian and the 5th British Infantry. The 1st British Parachute Division formed the assault force for Operation Slapstick, the assault on Taranto. Facing the Allies were approximately 130,000 Germans spread across the Italian Peninsula in eight divisions.[7]

Plans called for a D-Day assault force as well as convoys on D+2 and D+7 to land the initial force and its supplies. A convoy on D+19 would bring an additional seven days of supplies, and a D+24 convoy would bring fourteen days' worth of supplies.[8]

Only twenty-three days separated the cessation of combat on Sicily and the start of Avalanche. Because many assault craft had been engaged in the continued support of forces in Sicily, repair and refitting of these landing craft became the top priority. The limited time between operations prevented any full-size dress rehearsal between the army and the naval task forces. However, the divisions that participated in Avalanche had been able to conduct division-size landing exercises.

Many of the units within the Fifth Army lacked any experience in amphibious landings, so the Fifth Army had established an invasion-training center near Arzew, Algeria, in the spring of 1943. This center trained regimental combat teams and combat commands on the intricacies of amphibious operations, including beach operations. The capstone event was a practice landing operation of the 36th Infantry Division on August 26–27. The landings provided confidence to the participants; however, just as in the preparations for Husky, there was no practice of any large-scale supply off-loading.[9]

Mounting the Force

The mounting of the army in North Africa occurred mainly at the ports of Oran, Bizerte, and Tripoli. In Sicily, Fifth Army forces sailed out of the ports of Palermo, Termini, and Castellamare along the northern coast. The majority of VI Corps mounted from Oran, while most of the British X Corps mounted from Bizerte. Many of the units that had been involved in the mounting of Husky, in particular the US base sections, were also involved in the mounting of Avalanche. The 8th Embarkation Group of the Eastern BS, in particular, brought a wealth of experience. The recent practice from Husky helped make this second effort smoother because there was general agreement on the division of roles and responsibilities.[10]

Despite the fact that this was the third such type of assault, there were still major disconnects between the US Army and US Navy. The SOS issued a loading plan that had not been coordinated with the navy. The Mediterranean BS assumed control of directing the loading of ships at Oran, but not all of the base section's personnel had the expertise or knowledge necessary to load such vessels. Loading priorities changed, necessitating the loading, unloading, and reloading of some ships. Apparently, not all the lessons of the past made their way to those responsible for the mounting operation.[11]

Some of the boxes loaded on the assault craft sported incorrect labels through either negligence or intent. When a box marked "special equipment" broke open, out spilled piles of shoes. Others marked "medical supplies" actually contained coffee, sugar, and milk. In another incident, a box of incendiary bombs was stowed in an officer's stateroom, requiring later removal. Theater quartermasters worked to fill shortages in field-support units such as laundry units and bakeries. Planners forgot to requisition aviation fuel, causing additional scrambling of support units.[12]

Vessels were intentionally loaded beyond capacity—an accepted risk in view of the short voyage length. Transport vessels carried from 90 to 128 vehicles, and ships carried 300 to 800 tons of cargo. However, AFHQ did not publish the final allocation of shipping until just before D-Day, resulting in a continuous revision of loading plans.[13]

Initial Assault

The armada for Operation Avalanche was sizeable, including 642 ships and vessels as well as 925 ship-borne landing craft. Convoys sailed from Tripoli, Bizerte, Oran, Termini, and Castellamare, joining each other 12 miles off Salerno on the evening of September 8. Weather conditions were favorable for landings.[14]

The favorable beach gradients in the northern landing areas allowed the assault craft, including LSTs, to beach themselves directly on shore, thus eliminating the need for pontoons or causeways, which greatly decreased off-loading times. Ships approached the beaches with their ballast tanks dry, and once they were aground, pumped in 350 tons of seawater to hold them on the beach. The vessels on the southern beaches were not as fortunate, requiring pontoons to facilitate unloading.[15]

Plans called for supporting the force in three phases. For the preparation phase, the Fifth Army, the SOS, and the British Supply Agency would equip units and determine the initial support needs. As operations in Sicily were concluding, the Seventh Army was to turn excess equipment and supplies over to the Fifth Army. The second phase, estimated to last twelve days, called for invasion forces and supplies to flow over the beaches. The third phase would begin with the opening of the port of Naples.

British forces in particular had challenges during the preparation phase. A convoluted system of responsibility meant that British X Corps was under AFHQ for logistics planning but under the 15th Army Group for operational planning. Supplies in Sicily belonged to the 15th Army Group, supplies in Africa belonged to AFHQ, while supplies in the Middle East fell under British national control. This complicated system created a situation where there were few easy solutions to X Corps' logistics problems prior to the assault.[16]

Despite the positive experience of Husky in moving palletized supplies, the army did not have the time or resources to expand the effort for Avalanche, so most supplies were not palletized, thus requiring additional handling to unload and stack on the beaches. The navy's combat-loader vessels were slow to unload for a number of reasons, including poor loading in North Africa, congestion on the Salerno beaches, enemy fire, and an absence of the two-and-a-half-ton DUKW amphibious vehicles on the beaches.[17]

DUKWs had been included in the assault force, but, unlike for Sicily, the Fifth Army chose to use these vehicles mainly inland as trucks inshore of the beach dumps. In addition, the army dedicated 123 DUKWs to the hauling of 105-mm howitzers and ammunition for VI Corps. The navy viewed this reassignment as a disastrous diversion of resources: use of these amphibious vehicles in other capacities slowed vessel off-loading and exposed the DUKWs to loss through damage, attack, and maintenance problems. The impact of the absence of the DUKWs on the beaches was so great that Vice Admiral Hewitt recommended that for future assaults the DUKWs and their drivers should be under direct *naval* control until the unloading phase of the assault was complete. This was the only way to ensure that the ground forces could not divert the amphibious vehicles for other uses.[18]

Avalanche used a combination of US and British shore parties to operate the beaches. The British 56th and 46th Divisions used British Beach Groups, while the 36th Infantry Division was supported by the 531st Engineer Shore Regiment and a beach party from the 4th Beach Battalion of the US Navy.

The Fifth Army tried a new approach in the battle to organize the beaches. Instead of establishing an Engineer Special Brigade directly under the task force commander, as the Seventh Army had done in Sicily, the Fifth Army chose to use a subelement of the Engineer Special Brigade, the 531st Shore Regiment, for the beaches. This move was prudent because Avalanche had only one-third of the overall length of beach that Husky had. In addition, 6th Port, which had been responsible for the ports of Casablanca and Bizerte, had an early arrival date in Italy to establish control of the ports. Responsibilities for the support initially rested with VI Corps, which was later to transition to the Fifth Army.

Unlike some of the combat units, the 531st Shore Regiment contained a number of veterans from the landings in Africa and Sicily. The ninety officers and 1,980 enlisted men came under the command of Colonel Roland Brown. Brown and his men had been part of the regiment responsible for organizing the beaches around Gela for the 1st Infantry Division during Operation Husky. Here they were again, less than two months later, performing the same task for VI Corps, south of Salerno.

Two battalions of the 531st landed in the Gulf of Salerno, with one battalion held in reserve. Allied smoke screens proved to be so effective, however, that even the landing craft could not see the beaches, causing men and matériel to land on the wrong beaches, and the beach timetable soon became unattainable. Shore units made the best of the situation before them. One road platoon landed a mile and a half from their assigned beach but used the opportunity to build a beach exit road to get to the main highway. The landings were not going perfectly according to plan, but the plan merely served as a general guideline, anyway, once the elements of weather, time, and the enemy were factored in.[19]

By noon on D-Day, the 531st had finished building exits from the landing beaches to facilitate the arrival of forces and supplies. They could not establish shore dumps that day due to a lack of cargo trucks, but these vehicles arrived on D+1, and quartermaster units established supply dumps in the vicinity of Paestum, approximately 1.3 miles inland. By D+3, dumps had been established west of Highway 18, an important road that transversed the entire US line from north to south. A day later, the clearance of all supplies from the beach to the dumps was complete.[20]

Only one German division, the 16th Panzer, faced the Fifth Army at Salerno on September 9, but the German Tenth Army commander, General Heinrich Gottfried Otto Richard von Vietinghoff, quickly ordered the 26th Panzer and 29th Panzer Grenadier Divisions to reinforce the Salerno area from the south. He also ordered the 15th Panzer Grenadier and 16th Panzer Divisions to move to Salerno from the north. These two divisions totaled about 27,000 men but had only about thirty-seven tanks between them because they were in the process of refitting after evacuation from Sicily.[21]

The five German divisions immediately started toward Salerno, but supply problems hampered their movement. The Tenth Army did not have an organic quartermaster section so Kesselring's headquarters, Oberbefehlshaber Süd, had to handle all of the army's resupply. The results were less than satisfactory for the German divisions. There was no coordinated logistics plan, and the supply dumps did not necessarily match unit locations.

There was little fuel available from local Italian stocks. One panicked German officer demolished the petroleum-storage facilities and a coastal tanker at Sapri on the Gulf of Policastro when he incorrectly perceived that he was under attack. This loss of fuel, combined with the lack of a well-coordinated logistics plan, slowed down the movement of the 29th Panzer Grenadier Division, delaying its arrival at Salerno. Instead of arriving en masse, the division straggled in over the course of three days, preventing German commanders from fully committing the division while the US Fifth Army was still weak.[22]

Had sufficient fuel been available, the 29th certainly would have arrived at Salerno by the evening of September 9 and quite possibly could have made the difference in pushing the British X Corps off the beaches and dividing the Fifth Army. Back in the VI Corps area, the 531st proved that it was capable of doing more than running a beach or establishing a supply dump. As German snipers harassed the beaches at Paestum, Company D of the 531st and a company of infantry maneuvered to eliminate the snipers, destroy machine-gun nests, and drive-off German tanks that had hidden themselves in the city.[23]

Despite the seeming progress, the US Navy became increasingly concerned over the congestion of the beaches. Congested beaches slowed the unloading of the convoy, which could delay the scheduled return of the vessels. By D+1, Admiral Hewitt had seen enough and asked the army to dedicate 1,000 men to help clear the beaches. The Shore Regiment found as many men as possible, but the navy still estimated that by the end of the assault phase the navy had unloaded "only" 90 percent of the supplies onto the beaches. This was acceptable to the army, but the navy was clearly dissatisfied.[24]

The same problems seen at Sicily's beaches soon appeared at Salerno. From the perspective of the navy, the army had no apparent plan for beach operations. Supplies piled up on the beaches. Barracks bags and personal gear, hardly viewed as essential, landed "in huge quantities" as early as D+7. There was a shortage of personnel and trucks necessary to establish inshore dumps and to clear beaches. Beach operations virtually ceased at night and did not resume until the following morning. The proximity of Avalanche to Husky meant that there simply had been insufficient time to fix all the problems seen in the previous campaign.[25]

The arrival of tanks on the Avalanche beaches occurred much more quickly and smoothly than they had in Torch. The first thirty tanks were ashore on the southern beaches by 4:45 a.m. on the day of the assault. Supplies also arrived according to the landing plan, with more than 2,000 tons deposited on the beaches of Salerno by the end of D-Day. There were still problems concerning the sorting of items and clearing of the beaches, but at least the supplies were there.

By D+1, the 36th Infantry Division was occupying the high ridgelines to the north and south of their objective. German resistance was weakening, so by D+2 VI Corps had moved inland five to six miles from the beaches in all directions.

The Fifth Army quartermaster, Colonel Joseph Sullivan, arrived on the VI Corps Red Beach on D+2. Sullivan gives a slightly different account of the beaches than the one by the navy, describing them as "a marvelous sight. The dumps had been laid out, a wire road had been built and everything was proceeding in an orderly fashion." What explains the difference between the navy and army accounts? Possibly it is a matter of perspective: the navy wanted to see the beaches cleared as soon as possible so the ships could return to safer waters and adhere to the schedule, whereas the army merely wanted to see organization and progress. How one saw the situation depended on where one was standing and what one was hoping to achieve. Each service had a slightly different goal, which colored its assessment.[26]

The situation was similar on British X Corps beaches. Better beach gradients made off-loading of vessels easier, but there was a shortage of shore-party personnel and vehicles. Operations slowed as vehicles jammed the beach exits, and traffic jams developed along the limited road networks. Despite this difficulty, the British X Corps occupied Salerno by the morning of D+2, although the traffic congestion remained. Colonel Sullivan summarized the situation by stating, "Apparently, traffic control is not understood by British personnel as we know it."[27]

The most serious threat to the Allied beachhead came during September 12–14 as a strong German counterattack sought to split the US and British forces along the corps' boundary on the Sele River. This was the main fault with the invasion plan—it included a seven-mile gap between British X Corps and US VI Corps. Clark had called the gap "not too serious," but the Germans would use it to their advantage.[28]

This invasion was vastly different from that of Torch or Husky; the Fifth Army was facing a strong, determined, and capable opponent that could quickly reinforce itself. Even though the Germans were having trouble responding to the landings, they were able to amass enough combat power to split the Fifth Army in two. By September 12, the Germans were reinforcing at a rate equal to or greater than the Allies. Generaloberst von Vietinghoff sent a message to Kesselring's headquarters announcing that the US Fifth Army was on the precipice of annihilation.[29]

The main attack came midday on September 13. General der Panzertruppen Traugott Herr, commander of the LXXVI Panzer Corps, launched the 29th Panzer Grenadier and 16th Panzer Divisions along the Sele and Calore Rivers, hoping to take advantage of the boundary between the two Allied corps. Meanwhile, von Vietinghoff ordered the German XIV Panzer Corps to attack south of Eboli in an attempt to hasten the Allied withdrawal from the beaches. The German division made good gains, rolling over American defensive positions. Clark was unsure whether VI Corps could hold the beachhead and asked the Fifth Army staff to make plans for a possible reembarkment from the US beach—a scenario that would have been disastrous both for the operation and for the forces involved.[30]

AFHQ and the Fifth Army responded by focusing all available resources against the German attack. Fighters and strategic bombers converged on the German attackers. Naval gunfire was especially effective the nearer the enemy approached to the coastline. The third and equally important arm of the defense was the decisive response by Fifth Army units in dealing with the threat. All available men joined the defensive line. The Engineer Shore Regiment abandoned its beach operations and took a position on the VI Corps' right flank. Mechanics and truck drivers became infantrymen, an ad hoc change that strengthened the defensive line but limited the army's ability to distribute forces because it contributed to a lack of transportation. The small advance element of the base section, which had landed on D+2, even found itself on the line under the command of a noncommissioned officer. This was one disadvantage of landing early in an operation—the area was not yet secure.[31]

The crisis prompted AFHQ and the SOS to reinforce Salerno as quickly as possible. The Combined Chiefs gave Eisenhower approval to use eighteen LSTs that were transiting through the Mediterranean on their way to India. General Alexander directed that the rest of the 3rd Infantry Division replace service troops scheduled for landing in Salerno. Lieutenant General Clark considered sending the 82nd Airborne Division into Salerno, but there were insufficient landing craft and aircraft to move the entire division in the time required, so the 504th Parachute Infantry conducted a near flawless airborne operation that night.[32]

September 14 brought a new resolve to the Fifth Army beachhead. Allied air cover was provided by 187 B-25s, 166 B-26s, and 170 B-17s over the beaches of Salerno. Elements of the British 7th Division began arriving in the British X Corps sector. The last of the US 45th Infantry Division arrived, providing the Fifth Army with an operational reserve. The British admiral Andrew Browne Cunningham had ordered two cruisers and battleships from Malta and offered up another two battleships if needed. That evening, the 82nd Airborne's 505th Parachute Infantry Regiment landed just south of Paestum. With all of these elements combined, the Allies succeeded in making a massive reinforcement on, above, and off Salerno's beaches. Meanwhile, the assault force already on the beaches fought to hold its ground.[33]

On September 15, General von Vietinghoff and the German Tenth Army began to realize that they had lost the opportunity to push the US Fifth Army off Salerno's beaches. Allied air power made the daytime movement of forces and supplies dangerous. An attack by the Hermann Goring, 3rd, and 15th Panzer Grenadier Divisions made little headway against the reinforced British 46th Division. Late that day, the Tenth Army commander wired Kesselring for permission to break off the attack, then subsequently modified its strategy and assumed a defensive posture along the Italian Peninsula.[34]

The US Fifth Army was finally able to begin expanding the beachhead. The port of Salerno opened on D+4. The condition of the port was fair, and there was sufficient space to berth three coaster vessels and three LSTs within the port and to beach another twelve LSTs alongside the harbor. However, the port still fell within range of German artillery, so large-scale port operations could not begin until after D+15. From this point, the port of Salerno became the central site for off-loading ships supporting the advancing Allied army.[35]

In Sicily, the Allies faced a demoralized Italian force, which contained a few German divisions. For the attack of Italy itself, there were still questions about whether Italian forces would fight to defend the mainland, but an entire

German army with the ability to receive reinforcements directly from central Europe was also currently defending the area. Allied commanders needed to be able to deal with the threat—this meant that there had to be sufficient combat forces in the initial assault, or the Germans could push the Fifth Army off the beaches. The logistics of the invasion was a zero-sum game: more combat forces on the landing craft meant less service forces. The fact that many service troops landed on the beaches in the first convoys serves as a testament to the appreciation that combat commanders had developed regarding the role and contributions of service units.

The total amount of men and matériel landed over Salerno was quite impressive. Despite the challenges of the heaviest enemy resistance seen in the Mediterranean to date, inefficient beach parties, and occasional severe weather, the Fifth Army landed 202,066 men, 45,262 vehicles, and 153,930 tons of supplies from September 9 to October 8. As in Sicily, Salerno proved that a major port was not a prerequisite for landing and maintaining an army. However, the advantages of a large deep-water port were clear, and the army moved north to secure its main objective—Naples.[36]

Naples and the Establishment of the Support Base

An advance echelon of base section personnel landed at Salerno on D+2 and worked with Fifth Army staff in an advisory capacity. VI Corps had been responsible for support of the invasion force since D-Day, but the army headquarters had steadily grown in size, and by September 15, D+12, VI Corps was ready to relinquish the responsibility of sustainment to its higher headquarters. The line of communication was lengthening, and Lieutenant General Clark agreed that the time had come for the Fifth Army to take responsibility for the rear area, and so on September 15 its G4 Transportation Section became responsible for all beach operations and supply distribution.[37]

One lesson that the Fifth Army did note from Operation Husky was that SOS personnel needed to arrive early in the landing and establish a formal base section as quickly as possible. Once the army assumed control of the beaches, it had overall responsibility for the planning and oversight of administrative requirements, but it needed a capable headquarters that could conduct the daily management and execution of logistics.

Brigadier General Arthur Pence, former commander of the Eastern BS, arrived in Italy on October 2 as the commander of the 6665th Base Group (Provisional), later to become the Fifth Army Peninsula BS. Pence arrived in Naples along with the 6th Port headquarters and began establishing the main support base for US forces in Italy—the first base section on the European mainland.[38]

The citizens of Naples welcomed the arrival of the service forces. These units represented not only relief from German oppression but also a chance to rebuild lives, families, and the city itself. As in Sicily, life under occupation had been harsh. Conditions within the city of Naples were deplorable. Retreating German units had hidden large time bombs throughout the city, burned the city's stockpiles of coal, and torn up the port railways. Food was scarce, and there was a lack of clean water. The Germans had drained most of the city's main water supply, so there was only enough to last for eight days, and German forces still controlled the main reservoir. The people of Naples had been without bread for nearly ten days. When Pence and the service forces entered Naples, German artillery, positioned on the hills just north of the city, was still shelling the city.[39]

By October 5, base section personnel began occupying the Finanza Building in the center of Naples. Officers of the headquarters billeted in the Parco Hotel, while the enlisted men stayed in the west wing of the Naples post office. An explosion in the post office on October 7, however, forced the base section to relocate its troops to a former apartment building, and the Allies began a search of the city to find any remaining bombs. Meanwhile, the priority was to get the port into operation as quickly as possible.

The story of the rehabilitation and exploitation of the port of Naples is worthy of a volume of its own. This was a scene of notable dedication, skill, and ingenuity on the part of the Allied service forces to turn a destroyed port into one of the most important support facilities of the theater in a matter of days and weeks. Both the Germans and the Allies recognized that a major factor constraining the buildup and sustainment of any amphibious force was port capacity. The Allies worked to increase capacity as quickly as possible, while the Germans did everything in their power to damage and destroy port facilities and supporting infrastructure. Every major port seized in the Mediterranean, from Bizerte to Marseille, saw some degree of devastation.

The Allies expected that only a moderate amount of repair work would be required at Naples, but they encountered a scene of utter destruction on October 1. The Allies' strategic bombing campaign of the Mediterranean had included industrial transportation facilities within Italy, especially ports and rail yards. Thus, the ports of Naples and Genoa had been "thoroughly worked over" by the Allied bombers, which hurt the Germans but also would later hinder Allied use of these same facilities.[40]

In addition to the damage caused by the Allied bombing campaign, the Germans had taken the techniques used at Palermo and expanded them to bring a new level of damage to the port of Naples. For three weeks, German sappers had been hard at work destroying all usable equipment and sinking every available vessel in an attempt to render the port useless to the Allied force. Of seventy-two berths at Naples, only three were initially operable.[41]

Not only did the Germans sink vessels of all types alongside the piers and quays, but they sunk them in such a way as to make later clearance as difficult as possible. Bulkheads were blown apart to preclude rising by compressed air. Vessels lay on top of each other, and then German engineers added other debris such as mines and compressed gas bottles to the sunken mass. Explosives placed under gantry cranes damaged not only the cranes but quay walls as well. However, as inventive as the retreating German forces were at sinking ships and damaging port facilities, the Allied naval port party, under the command of the British rear admiral J. A. V. Morse, was equally as proficient at devising innovative measures of working through the clutter. Salvage teams quickly entered the port and began clearing the channel. Engineers left larger sunken vessels in place and built piers over them. Within two days, there were enough open berths to hold five Liberty ships and eight coasters; seventeen days later, British and US engineers opened an additional ten berths. Naples was on its way to being the hub of logistic activity for the Allies' Italian campaign.[42]

An advance echelon of 6th Port arrived on D+2 in Salerno, along with the 389th Port Battalion. The remainder of the port headquarters arrived October 1. Once in Naples, 6th Port went to work organizing the port. Constant rain hampered the discharge operations at Naples, just as it had at Casablanca. Men camped out in damaged buildings near the docks. Naples, however, was three times larger than the port of Casablanca and greater in capacity than the port of New York. The first day of operations saw 5,380 long tons unloaded at Naples; within six months, the total exceeded 2,375,000 tons—twice the peacetime discharge rate of the port of Naples and four times what the War Department had estimated.[43]

Port engineers quickly refurbished docks, quays, and gantry cranes. Within thirty days of its seizure, the port of Naples had sufficient throughput capability to support both the Fifth and Eighth Armies, a remarkable achievement. Even though the port of Naples was proving to be even more capable than originally expected, the Fifth Army worked to add additional ports to the growing support area. By the end of October, three smaller ports were working: Bagnoli, Pozzuoli, and Nisida. The Allies needed these additional ports to receive and stage the vast numbers of replacements and units that were landing on the Italian mainland, thus allowing Naples to focus on the reception of supplies and matériel.[44]

The original plan envisioned that all ports along western Italy were to be jointly shared between the US and British forces, but the British would have overall jurisdiction and command. This plan worked well immediately after the capture of Naples, but the relationship had to change as more and more US port forces began arriving in the area, and the volume of supplies arriving at the port

Figure 10. Sunken ship turned into pier (Military History Institute, Carlisle, PA)

were overwhelmingly for US troops. The British commanded the port of Naples until November 1, after which 6th Port had sole command.[45]

The actions of a single individual could create repercussions up and down the logistic chain. One such individual was Staff Sergeant Nick Orobello of Brooklyn, New York. Sergeant Orobello was determined to improve crane operations at Naples. An Italian speaker, Orobello headed into Naples and hired hundreds of local laborers, promising to feed them as well. The group loaded 8,396 long tons of coal into rail cars in seventy hours and cut the average time required to unload a coal ship from two to three weeks to three to four *days*. The faster off-load times meant that more supplies were flowing out of the port, and shipping was becoming more efficient. Word quickly spread throughout the city on the benefits of working at the port.[46]

The port soon proved to be a major employer for the local civilian populace as the civilian labor pool grew from 700 to more than 12,000. Besides providing these civilians with a reliable source of income, 6th Port also gave workers something that money could not always buy in Naples—food. The laborers spread word about the available meals, and port units soon found themselves feeding up to 20,000 civilian refugees a day in addition to the normal workforce. The ports and dumps became centers of activity and a key means of providing civil relief to the ravaged city.[47]

As the Allies struggled to get to Rome, German defenders worked to slow the Allied advance by any means possible. Retreating German units used every opportunity to remove or destroy critical components to communications and transportation facilities, which succeeded in slowing movement as well as diverting manpower and matériel away from the front. Complicating matters were the mountains, valleys, rivers, and ridgelines that characterize inland Italy.

Mountain Warfare

Whether one could characterize German troops as being the best in the world is debatable. Certainly, some US commanders felt that way when their attacks broke down and their armies could measure their progress only in terms of yards gained rather than miles. Perhaps the underlying explanation behind German effectiveness was the enemy's use of the mountainous terrain in Italy, which naturally suited the defender.

Rather than focus on the quality of the individual soldier, Field Marshal Kesselring envied the logistic capabilities of the Allied force, especially the Americans'. Not only did US forces enjoy an abundance of supply that allowed them to conduct large amphibious operations at any place of their choosing, but also the American equipment enjoyed a level of standardization that was missing in the German army. In addition, the Allied forces were motorized and able to rotate combat forces out of the front lines on a routine basis, something else the German army could not do because of a lack of units and a failure to modernize the entire army. These factors, the field marshal argued, were the main discriminators between the two sides.[48]

As the Fifth Army moved away from the beachhead, units increasingly had to operate in the mountains that paralleled the coastline. Motor vehicles could go only so far on the difficult slopes, where roads dwindled into paths. The typical chain of supply for a unit operating under such conditions was to transport cargo by truck to a truckhead. From there, supply specialists transferred supplies to Jeeps and drove them forward to company areas. In the company areas, the support platoons configured the supplies to fit on pack animals and moved

Figure 11. US Army pack train in Italy (US Army Quartermaster Museum, Fort Lee, VA)

them forward on mule trails. In extremely difficult terrain, units transferred supplies to a man-carried packboard for the final movement to the front lines. This chain was an effective, although not terribly efficient, means of supplying the frontline units, requiring time, labor, and multiple handling of supplies.[49]

The 3rd Infantry Division, having arrived from Sicily, was the only unit with its own pack train. Lieutenant General Clark became increasingly impressed with the capability provided by the mules and asked for recommendations on how to expand the capability to other units. The subsequent study from the staff recommended that the army procure an additional 1,300 animals.[50]

To meet the demand for service animals, the Peninsular BS established a US remount station in October at Persano, the site of the former Italian Remount Squadron. There were still about 200 head of usable stocks at the site, and the Fifth Army quartermaster quickly established the 1st Remount Station Headquarters to meet the demand for pack animals. Local mules were preferred because mules shipped from the United States tended to eat a different type of fodder. Just as US troops hated to subsist off their British cousins, the American mules detested Italian feed.[51]

The SOS bought animals according to a fixed range of prices in order to provide some level of standardization. The first rates in 1943 ranged from $80 to $150 per animal. By April 1944, prices had risen to a top limit of $250 for a mule and $300 for a horse. Each mule of Italian origin could carry about 100 pounds and traveled 20 to 150 miles per trip. The 513th Quartermaster Pack Company, a Black unit, was part of the supply effort. The 513th had two officers, seventy-five enlisted soldiers, and an allowance of 298 animals. The company had four platoons, each capable of carrying five tons of matériel per day, and was noted for its excellent efficiency. It saw service in both Italy and southern France later.[52]

American service forces initially operated the remount station, but the base section worked to transition as many jobs as possible to Italian military units, which freed US forces for other more-pressing duties and employed an Italian unit that had experience in working with animals. An Italian-speaking US officer was found and assigned to the operations section of the base section to serve as a liaison between the command and the Italian workers at the remount station.

The use of animals in the US Army was not a new concept, but Italy required pack animals on a scale not encountered in World War I. A typical US infantry division in 1917 contained only fifty-three pack mules. Most of the division's animals were for pulling wagons or riding, not for packing supplies to forward locations. As such, the older members of divisions in World War II did have some familiarity with working with animals, but there was little expertise on supporting battalions and companies by pack train.[53]

Supply of the pack units became a major task in and of itself. Just as trucks needed fuel and spare parts to run, the mules needed fodder, horseshoes, and saddles. The daily forage requirement could be as high as fourteen pounds of grain, sixteen pounds of hay, and one-eighth pound of salt per day for a large horse. The total forage requirement per month for the 15th Army Group equaled 1,125 tons of barley or oats and 1,500 tons of hay. The Italian army required an additional 300 tons of barley or oats and 600 tons of hay. The French army took care of its own needs.[54]

As the campaign continued, so grew the requirements for additional pack animals. In November 1943, the Fifth Army estimated that it needed 9,981 animals. By May 1944, the total Allied pack-animal requirement for Italy had grown to 18,236 mules and 3,023 horses, a total of 21,259 animals in twenty-three British, fifteen French, and six Italian remount units. In addition, the 10th Mountain Division, scheduled to arrive in November 1944, would bring another 6,000 animals. Forage would be a continuing issue for the 15th Army Group for the rest of the war.[55]

Figure 12. Packboards (US Army Quartermaster Museum, Fort Lee, VA)

The terrain was so rough that Allied planes were unable to airdrop supplies to the attacking forces because the supplies just rolled down the sides of the mountains. Likewise, there were no suitable roads to move supplies to the front lines, forcing units to evacuate the wounded by hand—a trip that took up to six hours. The troops at the front carried little in terms of rations or ammunition because they needed both hands to pull themselves up the hillside.[56]

To reach the extreme remote locations, such as the front lines at San Pietro, both mules and men had to carry supplies. Mules carried loads up to the end of a trail. From there, supply platoons transferred the supplies to men outfitted with backboards. These porters then carried the supplies up steep inclines, using ropes tied to trees as a means of leverage, on what could be a seven-hour round trip.[57]

The demand for packboards quickly exceeded available supplies. By the middle of January, the Fifth Army had ordered 7,000 packboards to help resupply remote outposts and small units. With a packboard, each man could carry 50 to 100 pounds of supplies. One company typically needed twenty men to carry a day's worth of supplies—ten carried rations, five carried water, and five carried ammunition. Support platoons packaged supplies into 25-pound containers so they could easily be transferred between mules and men.[58]

The threat from snipers or artillery discouraged fires along the front line, so it could be hard to make a cup of coffee, a staple in combat. To deal with this, cooks strapped three 155-mm brass shell casings to a packboard and filled them with boiling water. After the four-hour trek to the front line, the water arrived still hot enough to make coffee or soup—a welcome relief, especially during the winter months.[59]

The columnist Ernie Pyle arrived in the area during the start of the battle for San Pietro and observed the pack mules in action. He watched as soldiers unloaded the dead from the mules near a cowshed and then placed them on the ground. "Dead men had been coming down the mountain all evening, lashed onto the backs of mules. They came lying belly-down across the wooden pack saddles, their heads hanging down on the left side of the mule, their stiffened legs sticking awkwardly from the other side bobbing up and down as the mule walked." Pyle had little to say as he observed the off-loading of the dead, but the scene did inspire him to write one of his most famous columns of the war, "The Death of Captain Waskow."[60]

Much of the forage was available from local Italian sources, but 8,000 tons still needed to be imported. To help manage the collection effort, AFHQ formed a centralized board, the Joint Purchasing Forage Board, to collect and buy all available Mediterranean forage. Hay and grain had become as important to the fight as fuel and ammunition. To deal with the problems of caring for the animals, the army established a hospital for wounded pack animals at Persano on January 24.[61]

Training of the men assigned to the remount units was a continuing problem. Only half of them had ever worked with animals. Italian pack-mule units were often short of animals or men or both. Dark-colored animals were preferred because they were harder for the enemy to spot at night. However, as mules became scarce, light-colored animals increasingly found their ways into the ranks of the remount units. Troops sprayed the coats of these lighter animals with a 5 percent solution of potassium permanganate, an oxidizing agent that darkened the coats for up to two months at a time.[62]

Mountains and valleys were not the only challenges. The lack of a usable rail network meant that the service units had to rely more on truck transport to send the matériel of war to the front. As the distance between Naples and the front increased, so too did the demand for cargo trucks. The Germans' demolition of the railways had been more successful than expected, and so, as in North Africa, the flow of supplies in the early stages of the campaign depended on time and trucks. By October 21, the harbor of Naples was crowded with fifty-two ships, which had backed up awaiting discharge. The problem was not in getting equipment and supplies off the ships but in clearing it through the port and into the dumps. The crowding of ships not only represented a delay in off-loading needed supplies but also presented an inviting target for German aircraft.

On one occasion, German pilots managed to bomb a large pile of coal at one end of the port, igniting the mass into a large fire. At the opposite end of the port lay Mount Vesuvius, an active volcano. German pilots took advantage of the two highly visible landmarks by lining their aircraft up between the volcano and the burning coal to target ships anchored in the port. The only means to eliminate this target reference was for the base section either to move the burning coal or to cover it with a layer of dirt. After three days of work, the section managed to move part of the pile into the water and to pump a foot-thick layer of mud over the remainder, which removed one reference point, and after that the port experienced significantly fewer losses. The volcano was left untouched.[63]

General destruction of the rail lines, tunnels, and bridges slowed the establishment of rail service in Italy. German ingenuity had produced a giant steel hook that engineers attached to the back of a locomotive to plow up the wooden crossties. Demolitions cut the rails. The Germans destroyed everything they could not take with them. An advance element of the US 713th Railway Operating Battalion landed in Salerno shortly after the assault, but it took until the end of October to establish a rail link between Salerno and Naples with nine operational locomotives.[64]

Rail service gradually increased as engineers and rail personnel repaired or rebuilt tracks, bridges, and switches. The first task was to restore service around Naples. The rail lines surrounding the city suffered total destruction. Twenty-five major bridges were blown, and the Germans had damaged every rail, switch, and frog. Mines littered the rail lines and facilities.

The first locomotives put into service were cargo trucks outfitted with boxcar wheels. Arriving in the area on D+2, Allied rail troops located a serviceable rail line and several Italian boxcars, but no locomotive. Taking 2 two-and-a-half-ton trucks, they removed the traditional wheels and installed flanged railway wheels to provide the power to pull the railcars, demonstrating another example of American ingenuity. These truck/rail engines served as a temporary measure until the rail units could deploy heavy machinery from Sicily or North Africa onto the Italian mainland.[65]

Much of the repair problem concerned the fact that there were no stocks of general rail items available in Italy to repair the damaged rail lines. Workers had to use sidings to glean usable rails and switches. Rail workers had to clear demolished overhead bridges before repair of the tracks could begin. The Germans

Figure 13. Truck reconfigured as rail engine (US Army Center of Military History, Washington, DC)

were not solely responsible for the damage—Allied air raids had damaged the rail yards as much as the enemy's demolition work had.[66]

Engine operators, such as Sergeant Fred A. Tomer, were without a job until the rail lines were back in operation, so these men lent a hand wherever possible. Tomer decided to help out in the clearing of the rails by forming a track gang. The sergeant grabbed a private who had some track-gang experience, Private Alexander Parker. Together, they formed a twenty-one-man team that in a single day managed to reclaim eight rail cars and a section of track to place the cars on. The following day Tomer and his crew put an additional twelve cars into operation. Then, on the next day, the track gang repaired a German crane and built seventy-five yards of track. Sergeant Tomer was just trying to help and keep himself busy, yet it was exactly these kinds of endeavors that allowed the service units to bring the rear areas into operation as quickly as possible. Such efforts steadily increased the capacity of the Italian rail service. Allied rail troops eventually rehabilitated 6,233 miles of rail lines and 612 bridges across the country. In this war of attrition—conducted with aircraft and mechanized forces—the rail lines were the backbone of the ground-transportation network and allowed the support units to use the limited numbers of cargo trucks in the forward areas where the railways could not venture.[67]

Figure 14. Destroyed rail ties (US Army Center of Military History, Washington, DC)

By January 1944, the Railway Grand Division had reestablished service from Naples north to Vairano and east to Foggia. Its British counterpart had connected Taranto to Vasto. A rail line now connected the major elements of the 15th Army Group.[68]

In an effort to bring efficiency to the rail system, AFHQ placed all US and British rail units under the control of Brigadier General Carl Gray and the Military Railway Service. Working together with General Giovanni di Raimondo of the Italian military railway, Gray set the priorities and policies that governed rail reconstruction and organization throughout Italy. Cooperation was widespread, and rail service expanded throughout the war.[69]

To manage the limited Italian roads, rail lines, ports, and other infrastructure, the Allies formed the Italian Resources Commission. The senior British member of the commission was Major General Sir Brian Robertson, and the senior US member was Brigadier General Arthur Pence. The commission met regularly to set priorities and make allocations between the forces. Items adjudicated by the board included not only the expected matters of rail allocation but also such matters as shipping allocations for the British to bring in seed potatoes, the opening of Italian breweries, and the purchase of pack mules in Sardinia and Spain. Shipments of goods typically originated in North Africa before heading to Italy. Occasionally, the result of one decision could have unforeseen consequences on another.

On one such occasion, the action to open the breweries required a large purchase of hops. The SOS purchased the hops and, coincidently, shipped them to Italy aboard the same vessel containing a large number of mules. When the ship docked in Italy, the hundred-dollar-a-ton hops were gone, and the mules were bloated from the bounty of forage. Future shipments carried mules and hops in separate ships. Despite these types of incidents, the Italian Resources Commission process worked well, and the commission continued meeting throughout the war.[70]

Supporting the effort were a number of lumber mills and rock quarries. By the end of the war, Peninsular BS engineers were using sixty civilian mills to provide the lumber needed for all these projects, capable of producing 500,000 board feet of lumber per month, which were required for the construction of camps and bridges as well as the holds of Liberty ships. Each Liberty ship needed one million board feet of lumber to separate and store cargo in its holds.[71]

Unlike North Africa, Italy had a number of local resources that the Allies were able to purchase. Various items such as food, lumber, road-construction materials, iron, and electrical wire were available, which resulted in a savings of more than 400,000 tons of shipping space over the course of a year. The Allies' purchase of these supplies supported the Italian economy and meant that units did not have to wait as long to have requisitions filled.[72]

The Island BS established one of the largest medical centers of the war near Naples. It included a state-of-the-art laboratory at Bagnoli, which conducted advanced studies on hepatitis and native diseases. Italy also represented the first widespread use of whole-blood transfusions, penicillin, sulfa drugs, and evacuation of the wounded on aircraft. The Mostra fairgrounds contained three general hospitals, two station hospitals, and a medical supply depot. A large swimming pool facilitated physical therapy, while a 200-seat theater showed movies and live entertainment for the development of patient morale. By January 1, 1944, there were 14,992 hospital beds spread among thirty hospitals in Italy. The Allies had learned that patient survival rates increased the closer medical treatment facilities were to the front lines. Having a medical complex on the Italian Peninsula saved lives and lessened evacuation requirements.[73]

To stay abreast of the situation in Italy and to manage the administrative needs of the combined force, AFHQ established an advance administrative headquarters in Naples on November 1, known as FLAMBO—not an acronym but simply a code word for the forward headquarters. The main purpose of this headquarters was to coordinate administration and support between the Fifth

Army and the British Eighth Army in order to facilitate future operations against Germany.[74]

FLAMBO's role was to serve as a combined administrative staff and provide general administration of British forces. For the actual execution of administrative missions, the Peninsular BS handled support for all US units, while the British Number 2 District, stationed at Bari, provided for the local administration of British forces.

Up to November, the British Eighth Army had been receiving supplies from the port in Taranto. However, as the army advanced, the line of communication lengthened, thus increasing the transportation requirement for the service forces. Because the availability of trucks was an issue for the British as well, FLAMBO sought to reduce the line of communication by changing the port of supply from Taranto to Naples once the railway to Foggia was completed. Support of the armies was transitioning to Italy's western coast, while support for the air forces remained in the East. FLAMBO continued to coordinate general administration for British forces until February 1944, when it was absorbed into General Alexander's headquarters of the Allied Armies of Italy.[75]

Vive la France!

At the end of 1943, the French Expeditionary Corps began arriving in Italy, representing yet another demand on the US sustainment system. The 13,000 French forces included the 2nd Moroccan Division, the 4th Group of Moroccan Tabors, two mule companies, French service units, and 5,000 live sheep. The divisions of the French corps had been outfitted with US equipment in North Africa, and the French would fight under the Fifth Army, adding to the complexity of supporting that army. Planners had envisioned that the French would operate their own support base and staff it with French service forces. La Base 901 formed under the command of a Colonel Le Masle (no first name available) to serve as the main French support base. The Peninsular BS would provide supplies to Base 901, which in turn would supply the French corps. This system was relatively simple on paper, but differences in culture and expectations as well as a shortage of French service forces provided challenges that would carry over to the invasion of southern France.[76]

Providing support to the French was more complicated than it seemed at first. North African units were mostly Muslim, so a request for rations for a Muslim unit really meant rations with no pork, thus requiring additional handling and packaging at the ration dumps. French soldiers also tended to eat more than US troops, so ration issues did not last as long as intended. Units would be issued three days of rations and then eat them all in one day. Eventually, a system worked out where the French received 50 percent of their food needs from the US base section and the rest (wine, brandy, fish, and flour) came from French sources.[77]

Support of French forces was especially complicated because not only were there cultural and language differences to account for, but the accountants in Washington insisted that the SOS comply with the provisions of the Lend-Lease Act as well, adding a bureaucratic constraint to support. The War Department tried to institute a list of limitations on what the Fifth Army could and could not provide to the French, and policies required a full accounting of any items provided to the French so that the French might later reimburse the United States.

To further complicate matters, the theater imposed a requisition system for French forces that was cumbersome and unwieldly. Once in Italy, if the French needed an additional issuance of equipment, they were to submit a requisition to the Fifth Army G4. The Fifth Army G4 would then review the requisition and, if he felt it appropriate, forward the requisition to SOS NATOUSA. The French liaison office at SOS NATOUSA would send the requisition back to North Africa for the French General Staff G4 and Rearmament Section to review. Once reviewed, the requisition was sent to the Joint Rearmament Committee and then from there to NATOUSA for a final decision. This circuitous paper trail followed US War Department directions, which maintained that the Mediterranean theater was obligated to account for all transactions, but it was an unnecessarily complex and burdensome arrangement.[78]

By December, however, operational necessity trumped War Department directives. The theater adopted a local policy whereby the local base section could provide all necessary support to French forces, with the exception of major items of equipment. Major items, such as vehicles and weapons, required approval of the theater commander for issue. The US SOS became responsible for providing any supplies that the French could not provide for themselves.[79]

December also saw a tragic episode unfold at the port of Bari. On December 2, German aircraft attacked the port, including the US Liberty ship *John Harvey,* which was secretly carrying mustard gas artillery shells in its hold. The bombing of the *John Henry* produced an unintentional chemical attack on Allied troops who were either onboard the ship and other nearby ships or working to contain the damage. The lethal gas seeped into the air and water, contaminating the harbor and many of the men. The bombing at Bari produced more than 1,000 casualties, besides destroying the Adriatic Depot's headquarters. Fortunately, the abundance of Allied war matériel meant that the loss of twenty-five ships had only negligible impact on the Italian theater.[80]

On January 16, the War Department issued a new directive for all Italian forces operating under US command. Initial outfitting was to come from the French rearmament program. NATOUSA had the authority to supply anything that was not available from the formal rearmament process, and all items were accounted for using the French North African Lend-Lease account. Both US and French forces were to have the same levels of supply and supply tables. Consumable items, such as food, ammunition, medical supplies, and post exchange services were exempt from record keeping. This solution was much more workable for everyone concerned.[81]

Since the Free French forces had no industrial base to turn to, the burden of rearming French units represented a steady drain on stocks of Allied war matériel. In May 1944 alone, the French requested an additional 82,000 individual arms, 360 tanks, 6,000 machine guns, and 5,500 vehicles—in addition to those items already issued earlier in North Africa. This was the cost of incorporating into the coalition a force that could not provide for itself. The burden would only continue to grow as fall gave way to winter, and the Allies had to maintain their operations in a cold, harsh landscape.[82]

Winter 1943

The Allies had hoped that the Germans would withdraw to the Pisa–Rimini line north of Rome. By the fall of 1943, however, signs of German intention showed that Hitler planned instead to delay the Allied advance up the peninsula as much as possible in order to allow time for the German engineers to construct a formal network of successive prepared defensive lines. Winter would be a slow and grueling fight up the Italian Peninsula. This prediction came true as the fight bogged down due to poor weather, mountainous terrain, and a lack of landing craft needed to outflank the German defenses. As in Tunisia, circumstances, not the plan, dictated the pace of the campaign.[83]

As soon as it became evident that the drive to Rome would take longer than expected, the Fifth Army began looking ahead to identify requirements for winter clothing. Winter in North Africa had been cold and wet; winter in the mountains of Italy would be freezing cold and wet. Given the time needed to procure specialized equipment and clothing and to ship it from the States, the Fifth Army and the SOS had only a small window of opportunity to obtain what the units needed.

The initial requisition for winter clothing encountered the typical challenges of the military's supply system. The War Department cancelled a requisition for 100,000 sets of winter uniforms because factories no longer manufactured that particular type of clothing. Instead of substituting a different uniform of similar qualities, the War Department directed the Fifth Army to submit another requisition. The exchange of cotton uniforms for wool and vice versa occurred every six months, a process that required early planning and presented a large task for the supporting quartermaster units. Eventually, the Fifth Army submitted the right requisitions, and items such as winter uniforms, insulated tents, and heaters flowed into Italy.[84]

Winter also brought another challenge—illness. Two entirely different types of diseases affected two different populations during the first few months of 1944: typhus hit the civilian population in Naples, and venereal disease ran widespread throughout the Fifth Army. Both were a drain on the medical system, and each held the potential to influence the war effort.

As the local inhabitants returned to Naples, enterprising Italians saw an opportunity in the large numbers of Allied soldiers working or recovering in the Naples area. Licensed and unlicensed bordellos sprang up along an amusement belt of sorts, spreading gonorrhea and syphilis to unsuspecting troops. By the first week of January, venereal disease patients, mostly from the Fifth Army, filled 13–15 percent of the hospital beds in the Peninsular BS area. By the end of the month, 2,800 beds had been set aside solely for the treatment of venereal disease. It was becoming a problem that the Fifth Army could not ignore.[85]

To deal with the situation, the military police formed a vice patrol, commanders were held accountable for the numbers of cases in their units, and prophylactic stations containing condoms and washing materials were positioned in neighborhoods prone to prostitution. SOS medical units tested jailed prostitutes for venereal disease. If a woman tested positive, she received treatment before her release. Such measures collectively worked, and the incidence rate of these diseases gradually diminished.[86]

Typhus presented a great challenge, both in terms of the possible affected population and in prevention. A major typhus outbreak in Naples could quickly overrun the military's ability to treat everyone affected, limit the Allies' use of the port, and even possibly spread to the troop units. The Allies had seen the effects of typhus in past wars, and no one wanted to see it return.

A report by the Fifth Army surgeon provides some idea of the conditions that allowed diseases such as typhus to spread. Sanitation of the Italian populace was generally poor, and there were inadequate supplies of food, especially dairy products. Italians fertilized vegetable gardens with human excrement, and dairy cows frequently had tuberculosis or undulant fever. Drinking water frequently came directly from local streams. Typhus spread through body lice, which were common in the winter months, when many people stopped washing their clothes. The only good news was that the winter tended to kill off the mosquitoes, which spread malaria during the summer months.[87]

In early 1944, Naples was in a pre-epidemic state. Sanitation was still a concern, clean water was limited, and typhus fever was beginning to spread among the population. New cases developed at a rate of twenty to twenty-five a day through the month of February. In a quick response, the War Department deployed the US Typhus Commission under the leadership of Brigadier General Leon Fox, an army doctor and health officer, to Naples to deal with the threat.[88]

The commission worked with Italian authorities to treat the conditions causing the disease and to limit its spread. New insecticides, such as DDT (dichloro-diphenyl-trichloroethane), helped control insect populations. Within a month, the threat had passed, and no US soldier ever caught the disease. Commanders had a personal interest in keeping down the rate of disease—at any one time, soldiers with diseases or injuries occupied two-thirds of all hospital beds in the communications zone. The remaining one-third comprised battle casualties.[89]

Throughout November, rivers flooded, roads became impassable, and supplies rotted if they were not suitably packaged. Replacement soldiers were untrained in how to keep themselves dry, and young sergeants and lieutenants were untrained in the prevention of trench foot, the scourge of World War I.

Medical reports from the Fifth Army first began reporting cases of trench foot in mid-November. Before November ended, medics hospitalized 305 soldiers with the condition and reported another 1,323 cases in December. In the first ninety days of 1944, the Fifth Army had 4,000 additional trench foot cases, divided equally between the Cassino front and the Anzio beachhead.[90]

The British army had an effective means of preventing trench foot. Every unit conducted a daily foot drill in which a sergeant oversaw the men remove their shoes and socks, massage their feet, and then put on a dry pair of socks. This process stimulated circulation and kept feet dry. In addition, the British boots and socks were heavier than that of the Americans, so British soldiers were less susceptible to the disorder.[91]

The Fifth Army adopted this strategy and ordered an additional issue of socks for everyone in Italy. Trench foot was a medical problem, but one that the quartermaster had to solve. Socks became a priority item for issue, and, as with venereal disease, the Fifth Army held commanders accountable for the rate of trench foot within their units. These measures worked, and the incidence rate of trench foot in the army during 1944–1945 was only one-third that of the previous season.[92]

For the first time, the use of airdrops influenced major battles. US forces involved in the fight for Cassino during the third week of January 1943 received supplies by air for a five-day period during the fall of 1943. In 160 air sorties, supplies dropped in view of the enemy. In a testament of the skill of the air crews and the lessons that had been learned in packaging supplies for airdrop, 80 percent of the supplies dropped were received and in usable condition. Due to the difficult terrain, Cassino would continue to hold out until May 18, but the concept of airdrop for large units proved a viable option.[93]

One area of warfare in Italy that received little fanfare was the care of the dead. By the summer of 1943, units had become more accustomed to dealing with the fallen, and the Quartermaster Corps had refined the systems for establishing temporary cemeteries and dealing with personal effects. Unlike in Torch, where the dead were piled together on the beach awaiting a decision on what to do with them, graves-registration specialists accompanied the Avalanche landing force and quickly established a corps cemetery on four acres at Paestum, near the Gulf of Salerno. By D+4, two rows of graves were already complete. A personal-effects depot, run by a staff sergeant, removed the personal effects of each of the dead. Graves-registration specialists filled out duplicate forms—one copy accompanied the effects, and one copy remained at the depot. The depot cleaned all army equipment collected from the dead and returned it to a supply depot for later reissue.[94]

German POWs had the task of digging the graves, and the 36th Infantry Division chaplain interred the dead, who were buried without caskets. Blankets or mattress covers draped some of the bodies; some simply lay in their uniforms. Graves-registration soldiers used burlap straps to lower the dead into the graves.

The workload for graves-registration units varied, depending on the frontline activities. The record in Italy was set on May 30, 1944, when the US cemetery at Nettuno interred 245 bodies in a single day. Within the first three months of the campaign, the Fifth Army had established twenty cemeteries on the Italian mainland. Most were small, remaining open for only a few days and holding as few as twenty bodies. The trail of cemeteries was a witness to the flow of battle and the intensity of the fight.[95]

In the fall of 1943, it became clear that the Germans intended to fight south of Rome and to use the mountainous terrain to their advantage. Something needed to be done to break through or get around the German defense, and time was short—Allied assault craft were due to transit north to England to support the upcoming cross-channel assault. The Allies were looking for a shortcut.[96]

6

Operation Shingle

Anzio

> An audacious and enterprising formation of enemy troops . . . could have
> penetrated into the city of Rome itself without having to overcome any seri-
> ous opposition. . . . But the landed enemy forces lost time and hesitated.
> —General der Kavallerie Siegfried Westphal

By the middle of October, Allied hopes for a quick advance to Rome were fading. German forces provided a spirited defense along the Gustav line, running north–northeast from Cassino to Ortona. The Fifth Army hoped to use an amphibious assault along the western Italian coast as a means to bypass German defenses and force an enemy withdrawal. In the assault code-named Operation Shingle, the Allies hoped to land a force in the German rear that could drive east toward the Alban Hills. Commanders felt this would put enough pressure on the German defenders to allow the Fifth Army to break free of the line of defenses, link up with VI Corps at the Anzio beachhead, and then march on to Rome.

Vital to the operation was the use of sixty-eight of the ninety operational LSTs that remained in the Mediterranean. The Combined Chiefs of Staff had directed that AFHQ transfer these vessels to Europe in preparation for a cross-channel assault, but the possibility of securing Rome provided a powerful argument to allow the vessels to remain in the Mediterranean in support of the operation. The commanders in chief granted Eisenhower temporary loan of the LSTs for use in the Mediterranean, but only for a limited period. Without the LSTs, the Allies would be restricted to a series of frontal attacks against a strong, prepared defense.[1]

The competition for LSTs was fierce, and everyone had good reason to demand more. The SOS was using these craft to ferry supplies from North Africa to Italy. The Fifth Army wanted more LSTs in order to conduct amphibious assaults around German defenses. The War Department had promised the European theater headquarters in England two-thirds of all available LSTs by the middle of December 1943 to conduct landing training. In addition, the Pacific theater continued to demand additional assault craft as it prepared to begin the campaign to clear the Marshall Islands, scheduled for late January.

Landing craft were the key to make and sustain amphibious landings—landings were the hallmark of the war.[2]

The final decision for Anzio languished for several weeks as senior leaders tried to determine the best strategy for a stalled Italian campaign. Compounding the indecision was the movement of key personnel. On January 8, 1944, Eisenhower relinquished command of AFHQ to the British general Sir Henry Maitland "Jumbo" Wilson, commander of the British Middle East theater. Along with the change, Lieutenant General Jacob Devers arrived to serve as the deputy AFHQ commander and NATOUSA commanding general. General Montgomery was also to leave the Mediterranean for Europe to take command of the 21st Army Group. Generals Alexander and Clark remained in their current positions.

The combined change of supreme command in the Mediterranean and the extra attention from Churchill (who was recovering in North Africa from a case of pneumonia) led to renewed interest in the Anzio operation. On December 23, the prime minister became determined to seize Rome as quickly as possible before anyone could strip units and resources away from Italy for a possible operation in southern France. Logisticians questioned the feasibility of the Anzio plan because the LSTs were on loan only for a short period and could not provide long-term support of the landing force. The 15th Army Group and AFHQ noted these concerns but then commenced operations anyway because of a belief that the two forces could link up within eight to fifteen days.[3]

The assault force of 47,000 men and 5,200 vehicles sailed from Naples and surrounding ports on January 20 for the short voyage north along the coast. The number of ships and craft totaled 376.[4] Maneuvering around enemy minefields, the armada arrived off Anzio at five o'clock the following morning. The ground forces consisted of VI Corps under the command of Major General John Lucas, 3rd Infantry Division; a US Army Ranger force; and the British 1st Infantry Division. The initial force scheduled to land on D-Day held 27,000 US forces, 9,000 British forces, and 3,000 vehicles. The objective of this combined operation was to cut the German lines of communication and to threaten the forces opposing the Fifth Army.[5]

The landing area was just 30 miles south of Rome. The Anzio harbor was fairly small—it contained one jetty and was limited to vessels with less than a ten-foot draft. Only landing craft could navigate the shallow waters; Liberty ships had to anchor off the coast. The beaches and limited port facilities were not ideal for prolonged support of a landing operation, but commanders expected that there would be a quick linkup with Fifth Army forces soon after the landings. Unfortunately, the size of the Allied invasion force was determined by the number of available landing craft rather than by the operational objectives.[6]

The Allied nations divided support responsibilities for the force. The Peninsular BS handled support for all US units, while the British Number 2 District provided support for the British units. Once supplies reached the beach, they were the responsibility of the 540th Engineer Shore Regiment, which had a total beach party strength of approximately 4,200 men.[7]

To Advance or Not?

The attack began at 1:50 a.m. on January 22 as two British landing craft launched a barrage of 1,500 five-inch rockets on the beaches. There was no response; the landings were a complete surprise to the Germans. Allied aircraft flew more than 1,200 sorties that day as well, but there were few targets. The 3rd Infantry Division landed south of Anzio, and three battalions of Rangers occupied the town of Nettuno. The British landed on the beaches to the north to little opposition. An Irish Guards officer characterized the whole affair as "very gentlemanly, calm, and dignified."[8]

The landings occurred more or less as planned, although the British 1st Infantry Division had to make some adjustments. The presence of an offshore sandbar and poor beach gradients on the western beaches made landings difficult. Due to the poor beach conditions, the British closed their beaches and moved all additional landings to the Anzio harbor. Despite these changes, 90 percent of the men and equipment of the assault convoy arrived ashore by the end of D-Day, and most of the landing craft were returning to Naples. Casualties were light; VI Corps sustained only thirteen killed, ninety-seven wounded, and forty-four missing. Advantage on D-Day went to the Allies.[9]

At the end of D-Day, VI Corps had a favorable position. The Allied force made it ashore with minimal casualties; the 3rd Infantry and British 1st Division reached their assigned objectives; and the Germans were completely surprised. First Lieutenant John Cummings of the 36th Engineer Regiment was able to drive to the outskirts of Rome with his driver, where he observed German vehicles crossing the Tiber River. Eisenhower had given Lucas a deliberately vague order: to attack in the vicinity of Anzio, secure a beachhead, and "advance on Colli Laziali [Alban Hills]."[10]

The VI Corps commander, Major General Lucas, faced two options: push his forces as far forward as possible and hope that the bold move convinced the German Tenth Army to withdraw north or consolidate the beachhead and build up supplies to meet a possible German counterattack. Lucas chose the latter, perhaps in part because Clark had cautioned against an aggressive advance past the beachhead based on prior experiences at Salerno. Comments from Clark, such as "you can forget about this goddam Rome business," did little to inspire aggressiveness or risk taking. Despite limited enemy resistance, Lucas stopped his two divisions short of the Alban Hills and began to build Anzio into a base of operations, awaiting the arrival of the 45th Infantry Division. Some historians believe this pause on the beachhead was a missed opportunity; others note that it may have been a prudent move given that German reinforcements were already on the move. Axis commander in chief Albert Kesselring himself felt that the Allies had missed an opportunity to capture Rome.[11]

By January 29, VI Corps was ready to resume the offensive. However, the Germans had used the time to move the German Fourteenth Army Headquarters down from northern Italy to assume command of the beachhead. Both forces hurried to reinforce the area to gain an advantage over the other. The VI Corps attack on January 29 accomplished little. By then, there were an almost equal number of combat forces facing one another; VI Corps had 100,000 men on the beach, of which 25,000 were service forces. The Germans had 90,000 men, of which approximately 30,000 were involved in the support effort.[12]

Lucas's decision to stop the VI Corps advance short of the Alban Hills on D-Day and to wait in order to build up a base of operations before attacking became one of the most controversial topics of the war. Risk went along with either option: to advance or not to advance. Lucas later stood by his decision, explaining that going too far from the beach would have put his divisions in a position that was not sustainable if faced with a sizable German counterattack. Critics would claim that Lucas acted tentatively and missed an opportunity to move on Rome.

Lucas faced a particularly tough decision. The VI Corps experience at Salerno had shown how tenuous a beachhead could be when an invading force stuck its neck out. Lieutenant General Clark and the Fifth Army had come close to abandoning part of the beach because they could not adequately man the entire beachhead perimeter. Plus, venturing out to the Alban Hills extended the perimeter and possibly put divisions on the road to Rome without sufficient logistics support to sustain the drive.

The problem with Shingle was that the size for the attack force was based on the amount of available landing craft, not on the manpower needed for the mission. Politicians such as Churchill and high-level staff officers had more influence on developing the operational plan than did the commanders who eventually led the assault. As a result, the Allied force at Anzio had enough strength to remain on the beach but not enough to break through the German defense or sever the German line of communication with Cassino.[13]

Ultimately, Anzio degenerated into a four-month fight for land. Counterattack followed attack. VI Corps struggled to flow enough supplies into the beachhead to sustain the force. The Germans complained that they lacked the air cover needed to get close to the beach as Allied naval gunfire stopped German assaults as soon as the enemy came into range. Neither side was willing to give in, but neither side had sufficient strength to break the stalemate. German general Siegfried Westphal, Kesselring's chief of staff, wrote that "ammunition was always short. . . . Nor, of course, was there any superfluity of petrol, spare parts, or clothing. . . . After our inferiority in the air, it was these supply difficulties which caused the German leadership in Italy the most concern."[14]

While VI Corps focused its attention on dealing with elements of the thirteen German divisions that were responding to the invasion, supply operations progressed relatively smoothly at the harbor. Supply unloading moved more quickly than planned, and there were few, if any, supply or maintenance problems. Liberty ships were anchoring outside the harbor and transloading their cargo into landing craft, which could enter the harbor. Within a week, port units had fully off-loaded seven Liberty ships and 201 LSTs. The amount of men and matériel moving over the beach was sufficient to support a large offensive should the VI Corps commander decide to launch one.[15]

From the start of the planning effort, logisticians were not convinced that the two forces could link up in the advertised eight to fifteen days, so as a safety measure they allocated thirty-five days for shipment of supplies by sea. One convoy was to sail every three days to compensate for periods of poor weather that might affect off-loading operations at the beachhead.

An idea that had surfaced during the planning of the operation was to preload trucks at Naples and then drive these trucks onto assault craft for the trip to Anzio. Once at Anzio, the trucks could quickly drive off the vessels, which would shorten off-loading times and limit risk to the assault craft. Generals Clark and Lucas liked the idea; however, Churchill was opposed to it, thinking it was an inefficient use of trucks. Despite the prime minister's objections, the SOS made provisions to institute this ground–sea ferry service for sustainment of the Anzio beachhead.[16]

The combination of LSTs and two-and-a-half-ton cargo trucks rapidly became the centerpiece of sustainment for Anzio. Following the amphibious assault, follow-on convoys consisted of 14 LSTs and 500 combat-loaded cargo trucks, meaning that all of the trucks were preloaded with a predetermined breakout of essential supplies. Each truck contained 60 percent ammunition, 20 percent fuel, and 20 percent rations. The Peninsular BS loaded the vehicles at Bagnoli and staged them in serials at the port of Naples. The loaded vehicles then drove onto the LSTs, facing outward for quick off-loading, to make the trip to Anzio. Drivers remained with their vehicles.[17]

Once at Anzio, the vehicles simply drove off the landing craft at the beach and proceeded directly to the respective supply dumps. Each truck carried an average of five tons of supplies—twice its stated cargo capacity. Trucks that had arrived the day before had been loaded with salvage material, men on pass, and POWs and placed in a waiting line near the Anzio harbor. As soon as the loaded trucks exited the LST, the returning trucks pulled into the port and onto the vessel. To support this effort, theater logisticians made 2,000 trucks available in Italy to serve as a ground–sea ferry service to deliver supplies to the Anzio beachhead. For the Allies, this was a new means of resupplying a combat unit that maximized the capabilities of the equipment while minimizing the risk of enemy fire. Due to the immense truck requirement, this procedure could not support an army, but it could support a corps. This support of VI Corps continued until the Allied breakout occurred at the end of May 1944.[18]

One of the challenges facing the service forces at Anzio was the daily barrage of German artillery. The hills to the north and east provided German gunners with the visibility and range to target almost every dump or storage location. Following the initial assault, the Germans had quickly moved up some of the largest guns in their army. They used 372 artillery pieces at Anzio, of which 150 were larger than 105-mm. The two largest were nicknamed "Anzio Annie" and "Anzio Express." Each weighed 218 tons and fired a 280-mm projectile up to 36 miles. The beachhead was under constant bombardment by German artillery and the Luftwaffe. Ammunition dumps were particularly susceptible, with a single hit often setting off secondary explosions. Approximately 40,000 tons of munitions had to be stored under such conditions.[19]

The executive officer for the Fifth Army G4, Lieutenant Colonel Charles D'Orsa, deployed to Anzio as part of the Fifth Army advance element. D'Orsa provided a daily cable back to Naples, which helped convey the conditions of the beachhead. The conditions he described often resembled life at the front lines rather than in a support area. On February 29, more than a month after Allied forces landed at Anzio, German artillery still ranged the beach and harbor:

> The harbor and beaches were pretty hot today. It almost appeared as if they [German artillery] had observed fire. Shells were landing all over the Anzio docks and Yellow Beach area with air bursts over the area. They also registered on X-ray Beach with shells landing and also air bursts. Shells hit two LSTs today; one landed this morning about 10 o'clock on the loaded upper deck of one at the hard; destroyed two trucks, started a fire, caused about a dozen casualties. . . . Captain Haverty, of the Medical Section, who was supervising the loading, was killed.[20]

Unlike North Africa, Sicily, or Italy, there was no rear area for Anzio. Every inch of ground was within range of the enemy, which forced the service units to adapt how they performed their duties in order to protect themselves as well as their supplies.

The constrained beachhead meant that, as a means to limit risk, ordnance units could not disperse munitions—there was only so much available terrain. Logisticians had planned to expand the rear support area once a breakout occurred, but as time went by, that breakout became less of a probability. The beachhead was not expanding, but supplies and equipment kept arriving.

To deal with the situation, service forces, such as ordnance companies and detachments, became proficient at firefighting and using techniques, such as trench storage, to limit the exposure of supplies to enemy fire. Units formed trenches by using a bulldozer to push sand in the direction of the enemy, which provided a berm in front of the trench and restricted any loss to a single stack of crates. Unfortunately, the water table at Anzio was near the surface, and flooding of the trenches was an ongoing problem. Another solution was the use of L-shaped bunkers. Eight bulldozers could build sufficient bunkers in a day to hold 3,000 tons of munitions. This proved a very good design, and no fire ever spread from one bunker to another.[21]

If a bunker did catch fire, crews raced to deal with the problem before it could spread. Bulldozers piled dirt on top of trenches in an effort to smother the fire. If a fire had just started, hand equipment was often sufficient to quench the fire with water before any explosions occurred. In worst-case situations, commanders pulled their personnel back from the trenches and allowed them to return only after the fire had extinguished itself. On any given day, the beaches of Anzio lost an average of more than 31 tons of munitions to fires. However, the beach units' innovation, skill, and bravery prevented this ongoing problem from spreading too widely, which could ultimately influence the battle.[22]

For the first time in the war, there was no animosity between the frontline troops and the supporting service units. The shallow beachhead meant that everyone was sharing the same dangers, hardships, and frustrations. Soldiers called the evacuation hospital at Anzio "Hell's Half-Acre." The running joke was that German gunners were using the red cross on the tents as a target. The hospital was so dangerous that some of the injured would have preferred to stay in the forward areas rather than move to the beaches. However, evacuation to Naples meant that the wounded had to spend at least some time on the beach awaiting transport.[23]

Allied air and naval attacks on the German lines proved especially effective. Any German attack that neared the beach met concentrated gunfire from the Allied combat ships constantly cruising offshore. An attack by the German 715th Division came close to breaking though the British lines but failed due to the intense artillery fire and the naval guns of three Allied cruisers. On February

29, during one of the last German attacks at Anzio, 247 Allied fighter-bombers and 24 light bombers stalled the German advance. Field Marshal Kesselring later wrote that the Germans could have succeeded in pushing VI Corps off Anzio's beaches if it had not been for Allied air and naval fire support. Without this support, Kesselring argued, the Germans could have shut off the Allied supply line. Allied firepower not only stopped German attacks but also protected the lines of communication, which allowed VI Corps to survive.[24]

Support operations continued until adverse weather conditions shut down vessel unloading off the beach. High waves and winds simply made it impossible to transfer cargo from ships to the landing craft. The navy estimated that due to weather and high seas only two days out of seven would support unloading operations. In reality, good weather allowed vessel unloading an average of four days out of seven. Despite this greater average, some shortages did occur. In one instance, the munitions dumps at Anzio were down to six rounds of 105-mm ammunition. However, they did contain more than 100,000 rounds of tank ammunition (planned for a breakout that had not yet occurred). The tank rounds had about the same range as the howitzer rounds, so soldiers placed logs under the front of tanks along the entire perimeter, which increased the elevation of the tanks' main guns and increased their overall range to 14,000 yards. By doing so, 250 Sherman tanks performed the role of an entire division's worth of artillery and held the line until the weather subsided and ship offloading resumed. Yankee ingenuity had come through once again.[25]

By May 1944, VI Corps was ready to break out of the beachhead, and the Fifth Army was ready for the final push through the German winter line. More than 157,000 tons of supplies had been delivered to the beachhead. The Allies numbered some 90,000; German forces were down to 70,400. Strategic bombing raids had taken their toll on German supply lines and supporting infrastructure from Rome up to the Brenner Pass. German combat strength was ebbing along the front lines, while the Allies had managed to maintain a steady rate of resupply and replacements. In preparation for the breakout, more than 1,000 artillery rounds were stacked near each gun emplacement. An overwhelming barrage of fire would crack the line that had held the Allies for more than four months.[26]

A variety of factors had led to the prolonged stalemate. As noted earlier, the VI Corps commander, Major General Lucas, had failed to take full advantage of the surprise gained by the assault and intentionally did not push his divisions as far forward at Anzio as some felt that he should have. This is debatable. Lucas questioned whether he could sustain or reinforce those forces if they advanced too far from the beach. The historian Carlo D'Este writes that the only serious mistake Lucas made was to fail to capture the key terrain at Cisterna and Albano.[27]

Figure 15. Recovering the dead at Anzio (US Army Quartermaster Museum, Fort Lee, VA)

Part of the problem was also the reduced manpower of the British divisions—both at Anzio and all along the Fifth Army front. Due to the length of the war, the number of British casualties, and a relatively small population, British formations were at less than full strength, and sufficient replacements were not available to meet the demand. These units most likely lacked the power to punch a hole through the stiff German opposition.[28]

An additional factor to consider is that in January 1944 the US 36th Infantry Division had a shortage of equipment and limitations on its expenditure of artillery ammunition, which restricted the division's ability to cross the Rapido River as planned. The Fifth Army needed this crossing to help open up the Liri Valley, which in turn would open the way for the army to join forces with VI Corps at Anzio.[29]

The final factor was the weather. Torrential rains and a heavy snowmelt during January had flooded rivers. German engineers had diverted the Rapido River, creating a marsh that bogged down tanks and other vehicles. Movement was slow in both the forward and the rear areas.

In the end assessment, a lack of aggressiveness from the beachhead combined with poor mobility and a lack of combat power in several divisions contributed to the winter standoff.

Changes within the SOS and the Peninsular Base Section

While the divisions stalled at the front, a considerable amount of change was occurring in the theater support structure. The rear areas in Italy were growing and developing additional capacity every day. New units formed to meet the increased demands. Some changes occurred due to new commanders; others were due to developments in the tactical situation. The only constant in the communications zone was change.

Although the Peninsular BS's primary customer was the Fifth Army, other organizations also received their supplies and equipment from the base section or directly from North Africa. The Allied air forces, French forces, POWs, and civil populations put tremendous demands on the logistics system.

In November 1943, AFHQ reassessed its plans and assumptions based on the stiff German resistance encountered in Italy. Because Hitler was determined to oppose Allied occupation in Italy, Allied planners adjusted the basing of aircraft. To cover the offensive toward Rome, Eisenhower approved the buildup of six heavy-bombardment groups in Italy. These air groups belonged to the Strategic Air Force but would be a critical component in attacking German reinforcements, logistics, and rear area. Establishment of these bases in Italy would also support future operations in central Europe.[30]

By the spring of 1944, the number of US aircraft stationed in Italy, Corsica, and Sardinia totaled more than 6,500. The Fifteenth Air Force was operating from thirty-one different airfields, the Twelfth Air Force from twenty-eight. These numbers of aircraft and airfields represented a tremendous requirement for supplies, medical, transportation, ordnance, and engineer support—necessities that competed with the needs of the ground forces and were just as important.[31]

The basing of these groups had a significant impact on the operations of the Peninsular BS. Engineers and other service units had to refurbish or improve supporting infrastructure, such as roads, rail lines, and phone lines. The airfields needed large quantities of fuel, shipped by pipeline if possible or by truck if not. The transport of munitions would require truck and rail support. Fortunately, Naples provided sufficient port capacity to receive with the increased level of supplies, although the additional demands stressed the base station's ability to clear supplies from the port inland to dumps and airfields.

One of the strategic reasons behind the Italian campaign was the need to establish air bases on the peninsula that could provide the necessary range to hit targets in southern Europe. The area that provided the most suitable terrain for such a purpose lay on the eastern side of Italy, within the British area of responsibility.

North Africa had shown that British service units could not adequately support US forces. Each nation's systems, expectations, and equipment specifications were too different. In a novel approach, the SOS decided to form a service depot from base section units and attach it to the air corps. This depot would operate within the British sector to provide support to the Fifteenth Air Force, based in the Bari–Foggia area. This was an experiment with little guidance to work from, though.

AFHQ attached the Adriatic Base Depot Group to the Army Air Force Service Command effective October 17, 1943. The mission was simple—provide the air forces everything that they could not provide for themselves. Although the air corps was the main customer, the depot quickly became responsible for the sustainment of all US units operating in eastern Italy, including Major General "Wild Bill" Donovan's Office of Strategic Services operations with Yugoslavian partisans. The Adriatic Depot's main port was Bari, a facility shared with British forces on the eastern coastline of Italy, located at the intersection of the heel and boot of the peninsula. Supply depots were located at Foggia, Manduria, Spinazzola, Goia, Cerignola, and San Servero.[32]

By the spring of 1944, however, the experiment of attaching the Adriatic Depot to the air command had become intolerable. Experience showed that service units needed a formal assignment to the Adriatic Depot, not just an attachment for duty only. In addition, the depot needed a formal, direct relationship with the theater SOS headquarters in order to conduct business. The air force routinely downgraded SOS requisitions in priority, and its decree to flow all communications through the Fifteenth Air Force slowed down communications. The depot needed the same type of relationship that the Peninsular BS enjoyed with the SOS. AFHQ rectified this situation on March 15 by assigning all SOS service units in the area to the Adriatic Depot, and the Adriatic Depot began reporting directly to the theater SOS headquarters. This arrangement proved to be more practicable, and the Fifteenth Air Force, which was steadily growing to more than 120,000 personnel, still received its support.[33]

As the fight for the Mediterranean progressed, the theater continued to mature. Ports grew in capacity, and new ones came under Allied control. The air forces occupied more air fields, constantly extending the reach of friendly aircraft both in tactical and strategic support. The communications zone extended, increasing the demand for soldiers in the SOS. Requirements were once more outpacing assets. Something had to be done.

Improvement of the theater did not always mean adding more organizations. By 1944, the overlapping command-and-control structures of the US communications zone in the North African theater, and the SOS had become burdensome and inefficient. The division of responsibilities among support commands had never been quite resolved, and the overlapping headquarters represented more overhead than was required. The War Department simply could not fill all of the positions in all of the commands, as confirmed in an inspector general's special report. Two days after the report's submission, AFHQ was considering changes to the structure of the theater's support headquarters.[34]

Consolidation

In 1944, all the theaters of war felt the shortage of service personnel. The War Department Manpower Board called for an accounting of all SOS personnel, while the US headquarters in the Mediterranean created a policy that froze theater overhead. Any growth of the theater SOS headquarters would have to come from the base sections, the organizations that actually did the work.[35]

The shortage of service force personnel was limited not just to senior ranks. Eight officers from the War Department visited the Mediterranean in January 1944 to gain insights on how the 2nd Cavalry Division and other combat arms units, some 20,000 men, could best transition into service forces. The nation's priority had changed from filling combat units to filling the units that supported combat. The lessons of the Mediterranean were clear—an invasion force needed a capable support element, and in early 1944 the War Department did not have enough service forces for a cross-channel invasion.[36]

The issue of labor and the consolidation of responsibilities was a contentious topic within AFHQ. Major General Everett Hughes, the deputy theater commander and commander of the communications zone, felt that there should be no reason why the communication zone commands could not be consolidated under one officer. Others, such as the AFHQ G3, argued that commanding the theater meant more than logistics and that AFHQ needed several different commands to handle all the responsibilities.[37]

On February 24, 1944, Lieutenant General Devers settled the issue by abolishing the office of deputy theater commander. In addition, he combined the communications zone, theater headquarters, and SOS under a single commander, Major General Thomas Larkin. There were still two different staffs: a US theater staff operating as part of AFHQ, which set policy and priorities, and a SOS staff that coordinated the execution of support. However, for the first time a single general officer oversaw the activities of both, ensuring that their efforts were coordinated and that each was working from the same guidance.[38]

Larkin worked directly for Devers, the theater commander. The theater and SOS staffs as well as the base sections worked for Larkin. Thomas Larkin found himself one of the most powerful men in the Mediterranean. The new organization still used the nomenclature SOS NATOUSA (Services of Supply, North

African Theater of Operations, United States Army) until October 1944 to eliminate confusion. The name was later changed to COMZONE NATOUSA (Communications Zone, North African Theater of Operations, United States Army) effective October 1, 1944, and a month later to COMZONE MTOUSA (Communications Zone, Mediterranean Theater of Operations, United States Army) to match the name changes in the greater Allied theater. Having a single commander responsible for the US rear areas eliminated redundancy and confusion and represented a major improvement in the functioning and coordination of support.

No Longer a Priority Theater

Italy had the unfortunate distinction of being a campaign that started as a priority effort for the Allies but by late 1943 had become secondary to the upcoming cross-channel invasion into France. On December 25, Eisenhower received word that President Roosevelt had nominated him to serve as commanding general, Supreme Headquarters Allied Expeditionary Force (SHAEF). The new Allied force headquarters would plan and run the cross-channel invasion of France later in 1944. Eisenhower planned to take two senior officers with him: Omar Bradley and George Patton. Bradley had performed brilliantly in the Mediterranean and was destined for command of the US 12th Army Group. The War Department had held Patton in limbo in Sicily since August, still in command of Seventh Army but without any troops. Eisenhower recognized that Patton might still be an effective tactical commander and offered him command of an army, which Patton gracefully accepted. On January 1, 1944, Eisenhower departed the Mediterranean for his new command.[39]

The establishment of SHAEF also meant that the Mediterranean would soon lose those units and ships designated to participate in European operations. The 82nd Airborne and 1st Infantry Divisions, British XXX Corps, and convoys of LSTs departed the Mediterranean in early 1944 for refit and training in England. With their departure went the Mediterranean theater's ability to conduct large airborne and amphibious operations.

Lieutenant General Brehon Somervell, the commanding general of the War Department's SOS, visited Italy in November to see the supply situation firsthand. Somervell told supply officers that "the honeymoon for securing supplies is over for this theater [the Mediterranean]." The Fifth Army quartermaster, Colonel Joseph Sullivan, noted that "some other theater is to get most of the stuff." Planners, logisticians, and commanders throughout the Mediterranean would have to adjust their thinking and expectations to this new world order.[40]

The area most affected by the change in priorities was the availability of large-caliber ammunition. The combination of high expenditure rates and low supply meant that the theater had to impose restrictions on how many rounds each unit could shoot per day. The shortage was limited not just to US formations; British units felt it as well. By April 1944, the situation had become so restrictive that General Alexander visited London to make a direct appeal for an increase in ammunition resupply.[41]

The British Eighth Army requested 10 million rounds for the upcoming offensive to break the German defenses, but it received only 3.5 million rounds. Alexander was successful in convincing the British Chiefs of Staff that the 15th Army Group could seize Rome only if the forces in Italy received sufficient resources. An increase in munitions allotment had a direct impact on the Allied breakout a month later. Alexander noted that even with the shortages in ammunition, Churchill had still expected 15th Army Group to break through the German lines and secure Rome as quickly as possible. The prime minister never really accepted or at least never acknowledged to his generals that logistical considerations might delay offensive operations. Although it is possible that Churchill did not understand the impact of logistics on strategy, it is more probable that the prime minister's sometimes unrealistic expectations were just an act used to break through barriers and to get things moving along.[42]

The final breakout from Anzio started on May 23. The Germans tried to contain the breakout, but they had too few forces, little air power, and not enough support. By early June, the German defenses along the line from Ostia in the east to Pescara on the Adriatic began to crack, opening the path to Rome. On June 3, Hitler gave approval for a German retreat to the north.[43] Operation Shingle was a delayed success. It achieved its goals, but not within desired timelines. Shingle showed how a determined adversary could close off a beachhead and stymie an invasion force.[44]

For the Americans, the preceding eighteen months had provided a level of experience and knowledge that had professionalized the military. Equipment, systems, and processes had been refined to deal with the nuances of modern, mechanized warfare. However, the definitive test was about to start as focus shifted to the heart of Europe and to the bulk of the German defenses. The Allies were entering the phase of the conflict in which the war could not just be lost but won as well.

Part III

The Shift to Southern Europe

Master of the Craft

7

Operation Dragoon

Southern France

> The difficulties of supply eventually forced a halt upon us when we reached
> Germany, but the very rapidity of our advance across France had made that
> inevitable.
> —General Dwight D. Eisenhower

Although Allied operations in the Mediterranean were steadily progressing
throughout 1943, leaders continued to look for other strategic options as it
became increasingly clear that the Allies needed to open up a second front in
France to knock Germany out of the war. American planners had always looked
forward to making a landing operation in northern France, and now they
believed that the Allies could meet all the requirements for such an invasion and
still have sufficient resources available to conduct a secondary attack elsewhere
in France as well as to continue operations in Italy. This multifaceted approach
became the concept for the ending of the war in Europe.[1]

Allied leaders began discussing options for the second front at the Quebec
Conference in August 1943 and continued these discussions at the Sextant and
Tehran Conferences later that same year. Although the proposed amphibious
landings in northern France were a primary topic of the meetings, Franklin
Roosevelt, Winston Churchill, and Joseph Stalin agreed to include an additional
set of landings in southern France, which would give support to Operation
Overlord, the amphibious assault in Normandy, and help seal the fate of the
German army. In coordination with Overlord, the Allied leaders agreed to stage
a multidivision landing along the French southern coast, initially named Operation Anvil and later changed to Operation Dragoon. This would prove to be
one of the more important and debated decisions of the war.[2]

Mutually Exclusive Options

Although the Allies had agreed in principle to a second set of amphibious landings in France, the question of whether to invade France from the south as well as
from the north continued to be debated from the time of the idea's inception until

the actual event. Advocates such as Marshall and Eisenhower argued that a secondary invasion route into France would take pressure off the Allied armies landing in Normandy. Opponents such as Churchill and Field Marshal Alan Brooke argued that due to restrictions in available landing craft, the landings in southern France would come too late to help the Normandy effort and would serve only to divert troops away from the main effort in the Mediterranean—that is, the fight in Italy. Carrying on with the idea of a two-pronged landing meant limiting the 15th Army Group's ability to achieve a breakthrough north of Rome. Churchill instead advocated for landings in the eastern part of the Mediterranean that could support the fight in Italy. Both sides made convincing arguments, which produced a dilemma with no clear solution. The Americans favored a second set of landings in France, while the British maintained that the fight in Italy should be the priority of the effort in the Mediterranean. Stalin, favoring the more direct approach, put his support behind the American proposal.

The ideal solution would be to conduct a second set of landings simultaneous with those in Normandy. Once again, the physics of distance, space, and limited shipping came into play, preventing such an approach. The Allies would instead do the best they could with the Normandy landings in June and a supplemental attack in southern France in August.

While the debate raged on, planning for Overlord steadily continued. By the spring of 1944, that planning had been making progress for almost two years, and the Allies were now ready to initiate the operation. The pieces were set for the final chapters of the European campaign.

The US and British armies were not the only forces making gains in the summer of 1944. On June 22, the Soviet Red Army launched Operation Bagration—a multiarmy attack designed to retake Soviet territory and advance the front west toward Berlin. Within two months, the Soviet armies would decimate the German Army Group Center and clear eastern Poland of German occupation. Operation Bagration, conducted in coordination with the Allied attacks into France, forced Hitler's generals to react to Allied strategy.

On paper, the Allies held a sizeable advantage in terms of readiness and matériel, but physics determined the realm of the possible for Eisenhower's strategy. In the late summer of 1944, the beaches and ports of Normandy could not support the amounts of men and matériel programmed to flow into France. Eisenhower simply could not land all of the divisions and supplies as quickly as he needed them. The ports along the English Channel did not possess the throughput capacity needed to supply the three allied army groups that would eventually operate in France and the neighboring Low Countries.[3]

Port capacity would be a limiting factor throughout the rest of the war. For example, in the spring of 1945 the Allies were ready to cross over the Rhine River with sixty-eight available divisions. Atlantic ports under Allied control could support no more than thirty-five divisions, but the Mediterranean ports in southern France could support an additional thirty-five divisions. Thus, the Allies needed the ports of northern France *as well as* the ports of southern France to carry the offensive into Germany. Without this combination, the Allies would be able to support only half of the combat force they otherwise would have. The port at Marseilles ultimately handled more American supplies than any other port during the war, testimony to its vast capacity and strategic value.[4]

An additional consequence of invading France from the south is that it allowed Eisenhower to use the French First Army in its mother country. This gave French forces a direct hand in liberating France, adding to the size of the overall Allied offensive effort and providing an important public-relations aspect to the campaign. General de Gaulle remarked, "It is inadmissible for French troops at this stage to be used elsewhere than in France." The use of French troops was possible only with a third army group in France—the French would not serve under the command of either of the other Allied army groups operating in the North because the groups' commanders, Montgomery and Bradley, were Francophobes.[5]

Finally, the addition of a line of communication coming up from the South would reduce the risk of Allied failure. A single line of communication stretching from northern France might present a ripe target for the retreating but still viable German military. This would become alarmingly clear in December 1944 when Hitler launched an attack toward Antwerp with the goal of cutting off Allied supplies. The invasion of southern France would allow the Allies to land a third more combat divisions and associated support units. However, before operations could commence, Allied planners had the unenviable task of developing a feasible scheme of maneuver for two major operations, Overlord in Normandy and Anvil/Dragoon in southern France, while still supporting operations in Italy.

A Question of Strategy

When Italy removed itself from the war in September 1943, many expected a quick capture of Rome. Lieutenant General Clark and the Fifth Army instead found themselves in a slow, contested fight up the length of the Italian Peninsula through the spring of 1944. British generals, such as Alan Brooke and Jumbo Wilson, suggested that the strategic situation had changed and that the Allies should cancel Anvil in order to dedicate more resources to the fight in Italy. Anvil, they argued, required postponement until Italy fell. The British Chiefs of Staff in London concurred with this line of reasoning; however, the US Joint

Chiefs of Staff in Washington did not. And for military leaders in the Mediterranean, Anvil represented a rival operation that would only serve to deplete units and resources while providing little strategic advantage.

To keep Italy a priority within the Mediterranean, AFHQ needed to retain all of the divisions earmarked for Anvil, along with the associated support units, assault craft, merchant shipping, ammunition, fuel, and general supplies that enabled the divisions to fight. Accepting the British recommendation to keep Italy as a priority theater would, however, effectively preclude Operation Anvil. A major Allied victory in Italy might put the Allies in a better political position after the war due to the proximity of the Balkans, but this option would also introduce more risk into Allied military operations in France. Due to constraints in men and matériel, Allied leaders had to choose between two mutually exclusive options: conduct a second landing on southern France or continue to keep the Italian theater as a priority.

The stalemate at Anzio in January and February 1944 strengthened the British argument and led to a reassessment of the Allied strategy. The slowing of the Italian campaign and the vigorous defense offered by reinforcing German units forced planners to commit to the fight in Italy units and resources that had been earmarked for the invasion of southern France. Churchill recommended that the Allies continue with plans for Overlord but abandon the secondary sets of landings in France and instead concentrate their effort in Italy as a means to help the Overlord landings. The problem was that the Allies did not have the resources to conduct simultaneous operations in northern France, southern France, and Italy. Something had to give.[6]

The debate over Anvil raged throughout the spring. The British favored focusing efforts in Italy and canceling Anvil. The US Joint Chiefs of Staff, led by General Marshall, wanted to cease wasting resources in Italy, which they viewed as a secondary effort. In Washington, many planners recommended that the United States pull its forces out of Italy altogether and leave the theater to the British. The Joint Chiefs effectively canceled Operation Anvil in April 1944 when the United States failed to provide sufficient resources to prepare for the operation. The shortage of assault craft, especially LSTs, combined with the ongoing resupply efforts required at Anzio, prevented the reallocation of shipping resources for Anvil. Jumbo Wilson started planning a spring offensive in Italy, but he likewise lacked the amphibious shipping needed for the operation. Meanwhile, the Fifth Army moved ahead with its summer assault toward Rome.[7]

In a turn of events, however, the Fifth Army's summer offensive produced an unexpected breakout and the capture of Rome on June 5, 1944, and thus provided a new set of options for Allied leaders. The US and British alliance could revive Operation Anvil, which would turn Italy into a secondary effort, or they could focus their efforts on northern Italy. If they focused on Italy, the Allies could choose to move west into France or east into Hungary, but there would be no landings in France. Arguing that France had a more direct impact on the war, the Americans won the argument, and the British agreed to pursue the landings in southern France, although Churchill had not yet given up on his quest for the Balkans.[8]

Underlying the decision was Eisenhower's arguments for making a secondary set of landings in France. Initially, in 1943 and into the first half of 1944, Eisenhower and many of the American planners saw the landings as a way to take pressure off the forces landing at Normandy. If the Allies landed at two different locations, the Germans would have to focus on two distinct threats and would not be able to reposition reinforcing units already stationed in southern France. Although this was still true in June, Eisenhower modified his reasoning for the landings, now arguing that, perhaps even more importantly, he needed the additional port capacity that Marseilles and Toulon offered.

By June 17, 1944, Eisenhower realized that he needed additional French ports in order to allow for the rapid deployment of additional American divisions into the European continent. The advance in Normandy had bogged down, and the failure to capture any major port meant that the Allies needed to breathe new life into the operation in southern France in order to flow in the forty to fifty divisions awaiting deployment in the States and to provide sufficient port capabilities for sustainment of the additional forces. A massive storm hit Normandy ten days after the Allies landed there on June 6, wrecking the artificial piers, known as Mulberries, which were vital to bringing supplies ashore over the assault beaches. In a message to Jumbo Wilson and the Combined Chiefs, Eisenhower argued, "Our most important consideration is an additional port to be used in assisting the deployment of Divisions from the US. I consider vital the possession of another gateway into France." Now, more than ever, the Allies needed to land in southern France. By the middle of June, AFHQ was issuing orders for the transfer of units from Italy to Anvil, an important action because any unit coming out of combat needed ten weeks to rest, refit, and rearm.[9]

On July 2, the Combined Chiefs sent a message to General Jumbo Wilson directing him to be prepared to execute the landing in southern France on August 15. The purpose of the operation was to establish a Mediterranean bridgehead at Toulon and Marseilles and then to achieve exploitation toward Lyon and Vichy. The initial force list consisted of a three-division assault, followed by a rapid buildup of up to seven additional divisions. AFHQ tasked the SOS to prepare and load all of Anvil's combat units already positioned within the Mediterranean. An eyes-only message from General Marshall to Generals Devers and Eisenhower summarized the War Department's attitude toward Operation Anvil: "Your Theater is now functioning so that you will not be burdened with a great amount of administrative routine. . . . If the forces in Italy get bogged down on the Pisa–Rimini line, we should not long delay putting Fifth Army Divisions into the fight in Southern France. . . . The important thing is that we push Anvil to the utmost as the main effort in the Mediterranean."[10]

Marshall was clearly stating that, at least as far the United States was concerned, the upcoming invasion of southern France had become the top priority for the Mediterranean. Perhaps even more interesting is the stated belief that even though the Mediterranean would have to support two major operations simultaneously, Lieutenant General Devers and his 6th Army Group would not have to burden themselves with the details of coordinating supply, maintenance, and other administration. Marshall's message implied that the Mediterranean had indeed achieved a level of experience and capability among its support forces that would allow the army group commander to focus on the fight before him. Devers could not disregard the physical limitations of constrained resources, but he could leave the details of administration to other trusted officers, such as Tom Larkin.

Churchill was not happy with the decision to revive Anvil and even went so far as to recommend that the plan be modified to bypass the ports of Toulon and Marseilles and to land at ports in Brittany instead, believing that these ports would contribute more directly to the fight in Normandy. Alternatively, the prime minister argued further, the forces of Anvil could go to Italy to prosecute the campaign up the Italian Peninsula and eventually invade the Balkans via the Adriatic. The fight between Eisenhower and Churchill regarding Anvil's future went on for ten days.[11]

This argument showed the prime minister's apparent lack of appreciation for the challenges associated with his proposal. First, the Brittany ports were 1,600 miles from the loading ports in the Mediterranean, and many of the assault craft carrying troops could not handle such a voyage. In addition, the seas in Brittany were prone to storms and high tides. Third, the assault area was beyond the range of effective air cover. Finally, landing at Brittany would not support any follow-on operations beyond western France. Abandoning Anvil in favor of Italy and the Balkans meant exposing the forces in Normandy to an unprotected southern flank and limiting the number of Allied divisions and supplies along the Western Front. Roosevelt and Eisenhower dismissed Churchill's proposal out of hand.[12]

After the war, Eisenhower wrote that the purpose of an invasion of southern France was threefold: (1) to help liberate France; (2) to open an additional line of communication that would ensure rapid arrival of troops and provide sufficiency of supply; and (3) cut off German forces operating in southern Europe.

The reasons for the landings in southern France had changed, focusing more on logistics rather than on the enemy. However, with the on-again, off-again nature of the decision, planners had little continuity of effort. By July 2, though, with the final decision made, staffs at all levels went into high gear working to develop the plans needed to load, move, land, and support a major amphibious force for a landing date less than six weeks away.[13]

Planning the Invasion

The initial deliberate planning effort for Anvil had begun in January 1944 at the École Normale in Bouzareah, a suburb just north of Algiers. As with previous operations, AFHQ formed a planning group from the personnel of the army headquarters that would make the initial assault. In the case of Anvil, the Seventh Army provided the nucleus of the planning staff under the title of Force 163.

The Seventh Army had changed dramatically following its victory in Sicily. All of its divisions had transferred to other armies, either in Italy or in England. Patton was still in command but was due to rotate to England. In addition, seven of the eight primary members of Patton's staff planned to depart for England as well. To confuse the enemy, most of the army headquarters staff remained at Palermo while a small planning detachment under the lead of Brigadier General Garrison Davidson departed Sicily for Algeria.[14]

Included in the Anvil planning staff were eight officers from the Seventh Army G4, four from the SOS, and six from AFHQ. Over the course of a few weeks, this segment of the planning group established a second headquarters, named Rear Force 163, and stationed itself near the headquarters of SOS NATOUSA in Oran. This planning element produced the logistics plan for the invasion of southern France, which eventually grew to more than 300 pages.[15]

Expecting that the fighting in Italy would soon be over, the War Department selected Lieutenant General Clark to be the Seventh Army commander after Patton departed. In the spring of 1944, however, Clark was still fighting in Italy and had little time to devote to the planning effort. As such, the Force 163 planning team had to work with little oversight or guidance. Force 163 planned for a two- or three-division assault, followed by a sizeable buildup of up to ten divisions. The French were willing to put the French First Army under US command as long as the US provided all supplies, transportation, and other administrative needs. The plan assumed that no ports would be available until at least D+25. Until then, all supplies would come over the beaches.[16]

Command of French forces had been a touchy subject ever since Operation Torch. Britain and France had a history of stormy relations that influenced many senior officers' perceptions. The campaign in North Africa had shown both Montgomery and Bradley to be less than cordial toward the French. The French most probably would not have approved working for either of these commanders, and the two army group commanders would not have wanted to deal with the French anyway. However, the creation of an army group under General Devers avoided this predicament. Devers appreciated the additional forces, and the French longed for a chance to be a part of the attack that would free their homeland.

Working jointly, planners from both the US Army and the US Navy selected the Cavalaire–Saint-Tropez–Saint-Raphaël area as the best target to enable the capture of Marseilles and Toulon. These beaches provided good gradients for the landing craft and deep-water access for the larger ships. The beach defenses were not overly problematic, and there were suitable road networks leading inland. Once the Allied forces were ashore, the Argens River Valley provided the means to rapidly move west and cut off the two port areas from German reinforcements.[17]

By July, Eisenhower had identified four major objectives for the upcoming operation. First, Anvil was to contain and destroy any German forces that could potentially oppose Allied forces operating in northern France. Second, it would secure a major port to facilitate the landing of Allied reinforcing divisions. Third, forces would advance north to threaten the German southern flank and the German line of communication. Finally, Anvil would develop an Allied line of communication for support of advancing forces and support of the reinforcing divisions. Of the four objectives, three of them focused on issues of support.[18]

The plan for the landings was straightforward. The VI Corps, under command of Major General Lucian Truscott, would head the invasion. The 3rd Infantry Division would land on the left-most beaches near Cavalaire-sur Mer. The 45th Infantry Division would land in the center near Saint-Tropez. The 36th Infantry Division would land on the right, near Saint-Raphaël. The provisional 1st Airborne Task Force, which consisted of one British brigade and two US regiments, would drop into the areas beyond the beaches near Le Muy to augment VI Corps and the French 5th Division. The US VI Corps had a mission similar to that of previous landings: get ashore, establish the bridgehead, clear the ports, and drive inland. The Allies were determined to drive north as fast and far as circumstances would allow. The Seventh Army and the French First Army would follow VI Corps ashore, tailed by the 6th Army Group.[19]

Although senior leaders had been arguing about whether to proceed with the landings on southern France or not, the Mediterranean theater had in fact prepared itself to support such an operation. The SOS established the Northern BS in Corsica on January 1, 1944, under Colonel John Ratay to support all air

Figure 16. Brigadier General John Ratay (Military History Institute, Carlisle, PA)

operations originating from the island, but the SOS also tagged the Northern BS with conducting much of the loading of troops and supplies for the upcoming invasion of southern France.

John Ratay had enlisted as a private in 1914 and had served as an artilleryman in World War I. After the war, he had various assignments, including as military attaché assigned to Peking, China, and Bucharest, Romania. Prior to commanding the Northern BS, Ratay had served as commander of 20th Port and commander of the Atlantic BS, relieving Arthur Wilson. Eisenhower was impressed with Ratay's performance and recommended him for promotion to brigadier general in December 1943. Eisenhower wrote that Ratay "has proven a thoroughly competent commander, has continued to maintain a high state of discipline and has shown exceptional capacity in handling problems in connection with relations with the French and Arabs in Morocco."[20]

On July 6, NATOUSA activated the Coastal BS under the command of Major General Arthur Wilson, the former Western BS commander. Using insights gained from previous operations in which the support forces had arrived too late on the beaches and took too long to set up operations, Wilson insisted that all personnel assigned to the Coastal BS familiarize themselves with the details of the assault plan so they could assume responsibilities for supporting the army as soon as possible after landing. This time around, the base section would arrive early and take charge of the beaches as quickly as possible. In addition, when it became evident that the focus of the war in the Mediterranean was shifting north, AFHQ, NATOUSA, and SOS NATOUSA moved their headquarters from North Africa to Caserta, Italy.

Plans called for a phased buildup of supplies. Once the beaches in France were secure, five days of supplies would land every three days, which would allow support units to replace any consumed supplies as well as to slowly build up a theater reserve. Physics required the SOS to preload 100 ships with supplies in order to make the necessary timelines in France.[21]

Compounding the difficulty of the planning effort was the requirement to support not only all US forces attached to the 6th Army Group but the French First Army as well. The French First Army in Italy numbered some 75,000, but planners expected this number to double once the French divisions landed along the Riviera and had access to metropolitan French manpower. Under the command of General Jean de Lattre de Tassigny, much of the French First Army came from North Africa. The French had an organic support unit known as Base 901, but it had relatively few trained supply technicians or mechanics. From the beginning of the rearmament program, the French had focused efforts on producing combat units and had given the lowest priority to support units. According to US planning ratios, the French should have had 112,000 service forces to support their eight divisions, but they never fielded more than 29,000. Just prior to the invasion, the French First Army included only 12,500 service forces—10 percent of the total requirement. This weak service support forced Base 901 to form a close working relationship with the Coastal BS and the stationing of several French support liaison officers, although AFHQ prevented any formal liaison before July. Once in operation, however, the two service headquarters formed a close working relationship so beneficial to all involved that it lasted throughout the remainder of the war.[22]

The magnitude of the task of reequipping French units is hard to overstate. As of July 21, 1944, only 30 of 133 French units had at least a third of their authorized equipment. The Joint Rearmament Committee and the SOS worked to make up all possible shortages from supplies within the Mediterranean and from the French Lend-Lease account. In addition, the high percentage of Muslims in the French First Army meant that the SOS had to accommodate special dietary requirements in its combat provisions. The AFHQ G4 worked to coordinate unfilled requirements between assault units and the base sections.[23]

Early in 1944 the theater ordered all supplies for the upcoming operation, but then in April, when the future of Anvil appeared in doubt, the War Department canceled many of the requisitions. The resurrection of the operation in July forced supply officers to resubmit the orders, creating some shortages for any items that were not readily available. Luckily, on the belief that the Joint Chiefs might resurrect the operation, SOS NATOUSA had frozen the issuing of supplies identified for Anvil that were already in the Mediterranean. This foresight prevented the raiding of supplies by the Fifth Army in Italy and meant that the War Department had to ship only a limited number of priority items, such as ammunition, vehicles, and weapons systems from the United States. These items left New York in early July on ships instructed to arrive in Naples "with hatches open and booms slung so that the 6th Port could discharge the cargo without delay."[24]

To prepare the invasion force, the SOS had to reequip the 175,000 US troops pulled out of the Fifth Army as well as to outfit 150,000 French troops. In addition, the support units needed to build up enough supplies within the Mediterranean to support 450,000 troops for a thirty-day period. Any items not already available in the Mediterranean had to come from the United States.[25]

By August 1, planners had changed the operation's name from "Anvil" to "Dragoon" for security concerns, and units finalized their plans. The Force 163 staff moved to Naples and were absorbed into the Seventh Army headquarters. The plan called for a three-division assault initially under the command of VI Corps, with command later transitioning to the Seventh Army and then to the 6th Army Group. A force of 366,833 men and 56,051 vehicles were to be ashore on the beaches of southern France by D+30. To sustain this force, planners expected to move 277,696 tons of supplies over the beaches in the first month of operations.[26]

Once the landings began, each task force commander would become responsible for the resupply of his own troops. This responsibility would transition to the Beach Group Control Headquarters as soon as it was ashore and established.

Planners assumed that the Continental BS would be ashore and operational within five days of the capture of a port.

If all this sounded familiar, it was because by this point of the war amphibious assault planning had become more or less a matter of routine. Planning was still difficult and involved a great deal of coordination, but experienced staffs knew what questions to ask and what to expect. Systems and procedures that had worked well in previous operations, such as including support units and their equipment in early assault waves, now carried forward to the next operation; unsuccessful concepts did not. This is the main reason why Dragoon was able to occur as quickly and effectively as it did. Planning staffs, especially in the SOS, could now anticipate problems and develop the means to deal with them.

Planners never have all the facts they desire, so they rely on assumptions to fill in the holes of a plan, and Dragoon was no exception. Key assumptions for the support of Dragoon's forces included the projection that the first port, Toulon, would not be open until D+20 and could then handle 10,000 tons per day. Planners did not expect the port of Marseilles to open until D+40. Based on experiences in Sicily and Italy, the Allies fully expected both ports to have considerable damage from German sappers. Most importantly, the plan stipulated that the assault force would not be able to support itself more than 20 miles inland of the beaches until a port was operational. The key to Dragoon was the ports.[27]

Toulon had been a large French naval base, but the French had scuttled much of the fleet, so the port had a limited capability. Marseilles, however, offered a large discharge capacity that could support an entire army group. The location and capacity made this port a primary objective.

Marseilles was a major port with ten primary basins and 13 miles of quays. Its peacetime daily cargo discharge capacity was 20,000 tons per day, an important figure because the Allied planners expected that the 6th Army Group would need 15,000 tons of supplies each day. Connecting the port to the Rhône Valley was a series of roads, rail linkages, and canals.[28]

To open the ports, planners increased the size of the divisions. The US Army had learned the lessons of previous amphibious assaults and now worked to add support units to the assault formations in numbers that previously would have been unimaginable. For Dragoon, planners increased the part of each division not found within the headquarters or brigades and regiments from 25,000 to 45,000 men. In addition, each division's truck allotment increased from 4,000 to 8,000 vehicles. Planners and commanders alike were determined not to make the mistakes of the past, such as what happened in North Africa, where the lack of service units and vehicles had meant missed opportunities.[29]

Supporting the Assault

The SOS handled the loading of men, supplies, and equipment from bases in North Africa as well as in Italy. Naval ships and assault craft were generally loaded at Naples and Salerno, while merchant ships were loaded at Naples, Taranto, and Oran.

Naples, with its sizeable capacity, was the primary mounting port. As with the loading of forces for the assault into Italy, the base section formed an embarkation group to plan and conduct the loading of the ships. Oran served as the secondary mounting port and handled most of the French follow-on forces. With the leveraging of all the experience gained from past operations, all ships were loaded and ready to sail by August 8, right on schedule. To avoid excessive congestion in Naples, Vice Admiral Kent Hewitt decided to embark French troops already in Italy on the first follow-up convoys sailing from ports on the heel of Italy. The Fifth Army removed these units from the front lines and moved them to Taranto, and then the SOS reequipped them for Dragoon.[30]

Every convoy sailed on time with no reported major problems; experience was the hallmark of the Dragoon force. The Seventh Army had seen action in North Africa and Sicily. The 3rd Infantry Division had landed in Morocco, Sicily, and Salerno. The 45th Infantry Division had made landings at Sicily, Salerno, and Anzio. The 36th Infantry Division had made the initial assault at Salerno. All three divisions were part of VI Corps, commanded by the experienced former 3rd Infantry Division commander, Major General Lucian Truscott.

The naval task force, under Vice Admiral Hewitt, consisted of 902 ships and assault craft, along with 1,200 other craft carried aboard ship. The landings started in the morning hours of August 15 and went largely as planned. In total, 1,300 aircraft and 53 gunfire ships neutralized German defenses, allowing the assault to start at 8:00 a.m.[31]

The landing beaches spanned almost 60 miles across, making communications between the beaches difficult. For the first time, however, the navy provided a liaison officer to help coordinate the activities of the three naval beach battalions. Fortunately, the naval force encountered few underwater obstacles, so the landing craft generally had few problems delivering their loads.

Learning from past mistakes in planning support of amphibious operations, the Seventh Army G4 assigned port battalion crews against specific ships. These crews loaded the ships and then rode the same vessel to the assault area to conduct off-loading. Since the crews knew the contents of their ship and the location of specific items, off-loading occurred at new levels of efficiency, and essential items could be quickly located.[32]

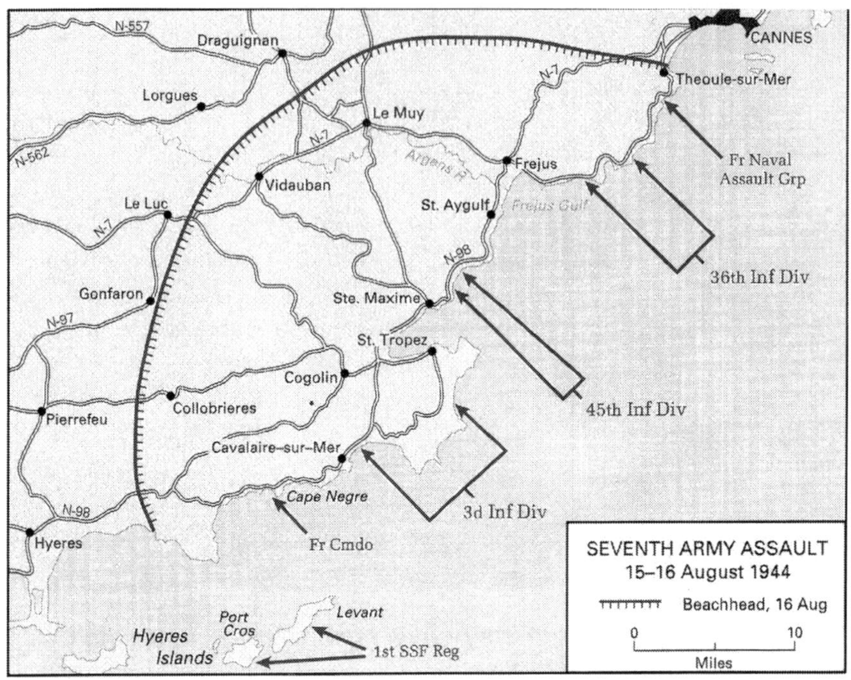

Map 5. Operation Dragoon, 1944 (Military History Institute, Carlisle, PA)

To help control the beaches, AFHQ and the navy established the central Beach Control Headquarters, which was designed to work through the Seventh Army G4. The army G4 controlled the off-load schedule, which prevented division commanders and their staffs from changing supply off-load priorities at the beaches—a problem experienced in past operations. The army G4 and Beach Control Headquarters called ships forward for unloading in accordance with the needs of the entire assault force, not just of one particular unit. An additional improvement was the marking of supply ships. Unlike earlier assaults, each supply vessel was clearly marked with the number corresponding to a master list of supplies and equipment. The army G4 called in ships based on demand, providing the ability to meet immediate needs. This proved to be a critical factor in adjusting to changing needs at the front lines as units achieved a breakout from the beaches sooner than expected.[33]

As in Sicily and Italy, engineer shore regiments landed with the assault elements to control the beaches and establish the initial supply dumps. For this operation, however, the shore regiments did not retain this responsibility for long. On September 9, the Continental BS assumed control of the beaches and dumps, allowing the engineers to revert to army control for use on more-traditional engineering projects.[34]

Initially, planners had expected to fight up to fourteen German divisions in southern France, but three of the enemy divisions had moved north to defend against the Normandy invasion. Instead of Dragoon pulling German divisions away from the North, as originally planned in 1943, the timing resulted in Overlord pulling divisions away from Dragoon. To further help matters, only three enemy divisions actively opposed the Dragoon landings because of the expansive coastlines the Germans were trying to defend. The result was a general lack of strong resistance, which allowed the Dragoon forces to move inland much farther and more quickly than planned. This quick movement equated to a tactical coup, but one that placed greater and greater demands on a support structure that was struggling to support ongoing combat while also establishing itself ashore.[35]

As planned, the 3rd Infantry Division landed to the southwest of Saint-Tropez along the beaches in the Bay de Cavalaire and the Bay de Pampelonne. The division quickly secured the left flank of the invasion and then raced inland to capture Saint-Tropez and link up with the 45th Infantry Division, thus securing the Saint-Tropez peninsula by dark on the first day of the landings.[36]

The 45th Infantry occupied the center of the assault formation, landing on the beaches running from Cape Sardineau to Point Alexandre along the Gulf of Saint-Tropez. Steep cliffs arose from the back of the beaches, and only a single road served as an exit. Resistance was light because air and naval bombardment had already destroyed much of the German defenses.

The 36th Infantry Division landed along a stretch of beach ranging from the mouth of the Argens River north to Antheor Cove. This area held the majority of the German defensive units. The landings started on time and initially met with light resistance. By late morning, however, parts of the division encountered firm opposition, delaying the advance inshore. Strong defensive lines protected Saint-Raphaël, including mines, booby traps, antitank ditches, and artillery. A determined German defense along the 36th Division's "Camel Red" beach forced the 142nd Infantry Regiment to land at a different beach, thus delaying the regiment's arrival onto shore by six or seven hours. Regardless, the 36th Infantry Division achieved its objectives, and by the afternoon of August 16, D+1, VI Corps had advanced farther inland than planners had ever thought possible.[37]

The French First Army began landing on the beaches on D+1. Four divisions quickly made it ashore and then began working their way to the west, reaching Toulon on August 22. The ports of Marseilles and Toulon surrendered to French forces on August 28.

The quick drive inland produced an almost immediate shortage of fuel. Supply planners had expected a hard fight on the beaches, so almost three-quarters of the supplies accompanying the assault force consisted of ammunition, and very little fuel was loaded onto the first waves of assault craft. The shortage of fuel on D+1 forced a revision in unloading priority within VI Corps. Logisticians diverted DUKWs and LCTs to a merchant vessel carrying 50,000 gallons of fuel. Beach capacity was not an issue, but operations had deviated from the plan. By D+4, units were more than 100 miles from the beach, invalidating the assumption that operations would be limited to 20 miles from the beach until the base section opened a port. The three US divisions were using more than 100,000 gallons of fuel per day, but as of August 21 only 11,000 gallons remained in beach depots. This shortage forced the Seventh Army to prioritize the off-loading of fuel and to restrict fuel consumption.[38]

By August 25, fuel was becoming a deciding factor for both sides. The US 36th Infantry Division and German 11th Panzer Division had arrived around Montelimar, and each sought to destroy the other. Fighting continued for forty-eight hours, but each side found itself hamstrung by shortages in fuel, ammunition, and available roads. Neither was able to destroy the other. Both sides penetrated the other's line at various points, but neither was able to exploit the situation because each had become too weak. The 3rd Infantry Division had moved forward by this time and sought to pursue the retreating Germans but also found itself slowed by persistent fuel shortages. The lack of supplies cost each side an opportunity: the Americans lost an opportunity to destroy a German army, and the Germans lost the opportunity to eliminate one, if not two, US divisions.[39]

To remedy the fuel shortage, the Seventh Army used the trucks of newly arriving units to move fuel to the forward areas. In addition, the army opened a rail link between Frejus and Saint-Maximin, leading to the establishment of a forward supply dump for fuel, rations, and ammunition. An additional shipment of 6 million gallons of fuel arrived off the French coast from Italy on August 28, which was good news, but it created the challenge of moving the fuel from the beaches to the forward units. VI Corps created three provisional truck companies to cover the 213 miles from the division areas to the fuel dump. The Seventh Army opened additional dumps on September 2 at Bourgoin, La Tour-du-Pin, and Montelimar, which reduced travel distance to between 24 and 83 miles. Even this opening provided only limited relief, however, because units kept moving farther into the interior, forcing quartermasters to push dumps forward. The pressure to deliver fuel forward was so great that VI Corps had to restrict the truck drivers to driving no more than seventeen hours a day in order to prevent their exhaustion. Fuel and ammunition would remain in short supply until the Allies could put the ports into operation, repair the roads and rail lines, and run a fuel pipeline northward from the coast.[40]

Within a week of the initial landing, the navy realized that it needed an additional means of delivering supplies to the Seventh Army, one that bypassed the beaches. On D+10, Admiral Hewitt issued orders to open Port de Bouc as a means of providing this additional capacity. Port de Bouc was located west of Marseille and close to the mouth of the Rhône River. Besides containing the usual port facilities, Port de Bouc also held a sizeable capacity for receiving and storing fuel. Engineers started clearing the port at the end of August, and within ten days the port was in operation. The nearby Arles Canal allowed supplies to be ferried 30 miles inland. Port de Bous served as the initial discharge port for all of Dragoon's forces and later developed into the primary discharge point for air force fuel and other supplies.

The Southern Ports—Underwriting Victory

Both Toulon and Marseilles fell to the Allies on D+13, allowing army and navy engineers to begin clearing the ports and to repair facilities. As expected, both ports had been badly damaged by friendly bombing and enemy demolition. The navy focused its efforts on Toulon, while the army handled Marseilles. German engineers had destroyed most of the buildings and roads. Scuttled ships blocked berthing spaces and the harbors. Using the lessons gained from previous efforts, such as at Palermo and Naples, the Allies quickly rehabilitated the two ports. The use of prisoner labor and French arsenal workers greatly aided the effort.

The destruction at Marseilles was thorough and efficient. Only one of the port's twenty-three piers was serviceable. There were no operational cranes or other discharge equipment. Mines had been scattered throughout the port and in the harbor, including a marine mine that was set to detonate sixty-three days after the Germans retreated. The retreating Germans damaged or blocked all piers. The damage at Marseilles was worse than the damage at Naples. The enemy had noticed how rapidly the Allies had rehabilitated Naples, so at Marseilles the Germans not only scuttled vessels but also did so in such a manner as to make clearing the vessels as difficult as possible, using explosives to break ships' backs before the ships sank. Vessels lay up to three layers deep in the channels and at the piers. Thousands of mines rested among the wreckage. The Germans had sunk seven ships at the western entrance of the harbor and had scuttled another sixty-five. In total, the Germans had sunk more than 200,000 tons of ships, including four large ships that now lay across the main channel. The ships sunk alongside the quays lay in such a way that engineers could not build piers over them, as had been done in Naples. By September 1, however, the first three Liberty ships entered the port and were discharged using DUKWs.[41]

Although planners had not expected to clear the Toulon and Marseille ports until D+20 and D+40, respectively, by D+14 both ports were in some stage of operation. The Germans had become adept at damaging ports, but the Allied engineers and port units had become equally proficient at rehabilitating them and their facilities. The first discharges of cargo at a pier in Marseilles occurred on September 8, totaling a mere 1,219 tons. However, within a month after the landings, the ports could handle any demands the Allies made upon them. By September 15, the ports were sending up to 10,000 tons of supplies up the Rhône River each day. By mid-October, the port of Marseilles employed 11,000 men and discharged more than 19,000 tons of supplies and equipment per day. In comparison, the port of Cherbourg, Eisenhower's main port for supplying the armies in northern France during the summer and fall of 1944, could handle only 7,600 tons per day two months after its capture.[42]

The planners for Overlord had expected to use the ports of Cherbourg, Brest, and Antwerp to supply the drive across France; however, this expectation was not realized. The First Army did not capture Cherbourg until June 26, 1944, twenty days after the initial Normandy landings. Patton failed to put a priority on the capture of Brest, and by the time the port fell on September 19, the army had advanced too far to the east, and the Germans had damaged the port to such an extent that it was of little use. Antwerp proved to be the biggest challenge—Montgomery's forces captured the port on September 4, 1944, but did not clear its estuary approaches of defending Germans until November 26, 1944. This meant that at the time of the Normandy breakout, the only operating ports were Cherbourg and the British beaches, with Cherbourg not reaching full capacity until the middle of August. As a consequence, the European theater lacked the necessary port capacities in northwestern Europe to support a multi-army pursuit across France during the last half of 1944.[43]

Due to the competing demands for shipping from around the world, the Mediterranean had to use all available transportation resources as efficiently as possible. The SOS had to off-load convoys as quickly as possible, so ships could conduct faster turnarounds. At Marseilles, a seventeen-vessel convoy was unloaded in only seven days and seven hours, providing 82,017 tons of supplies and equipment for the fight. A second convoy of sixteen ships discharged 89,575 tons in only six days and eighteen hours. From the opening of Marseilles on September 8, 1944, until January 25, 1945, the port discharged more than 2 million tons. This rate of discharge represented a transformational difference from the low rate achieved by the port units that had worked so ineffectively to clear the ports of North Africa two years earlier. Had the Allies chosen to land in France in 1943, they would not have had these same skills and abilities.[44]

The three southern ports, combined with a distribution network, provided the capacity that Eisenhower had sought. By the end of September, Marseilles could handle sixteen Liberty ships and forty-five LCT landing craft. Toulon could unload nine Liberty ships and thirty-one LCTs, while Port de Bouc had room for three Liberty ships and one tanker. Added together, these ports' discharge capacities exceeded the resupply rate required for the advancing armies. However, port discharge rates meant little if there was not a corresponding road and rail network, including sufficient locomotives, rail cars, trucks, and operators to move the forces and matériel forward.[45]

Race to the Rhine—the Seventh Army Moves North

The successful landings and the lack of a coordinated German defense presented an opportunity for the Seventh Army. Unlike in southern Italy, retreating German units in France did not possess the strength to delay the Allied advance for any protracted period. German engineers were, however, able to destroy or sabotage critical infrastructure leading north, such as roads, aqueducts, bridges, and rail lines. Nonetheless, Allied experience in dealing with this type of obstruction meant that army and navy engineers could repair damaged transportation networks as quickly as possible, although that effort did not always keep pace with the advancing units.

The first six weeks of Operation Dragoon consisted of a relatively uncontested assault, followed by a 400-mile pursuit up the Rhône Valley. Taking advantage of the situation, the Seventh Army commander, Lieutenant General Alexander Patch, urged his division commanders to move inland as quickly as possible in order to deny retreating German forces the opportunity to establish a prepared defense. While the Allies were moving up into southern France, Patton's Third Army was pursuing its own breakout from Normandy and was soon approaching Paris. Montgomery was driving into northern France. Each force needed uninterrupted supplies of fuel to keep moving, which put significant pressure on the service units supporting each army to keep pace with the combat forces.[46]

Ramifications of Unexpected Success

At the end of August, D+15, combat formations in southern France were on a line of advance that had not been programmed for capture until D+60, but there were only sufficient service forces in southern France to support a quarter of

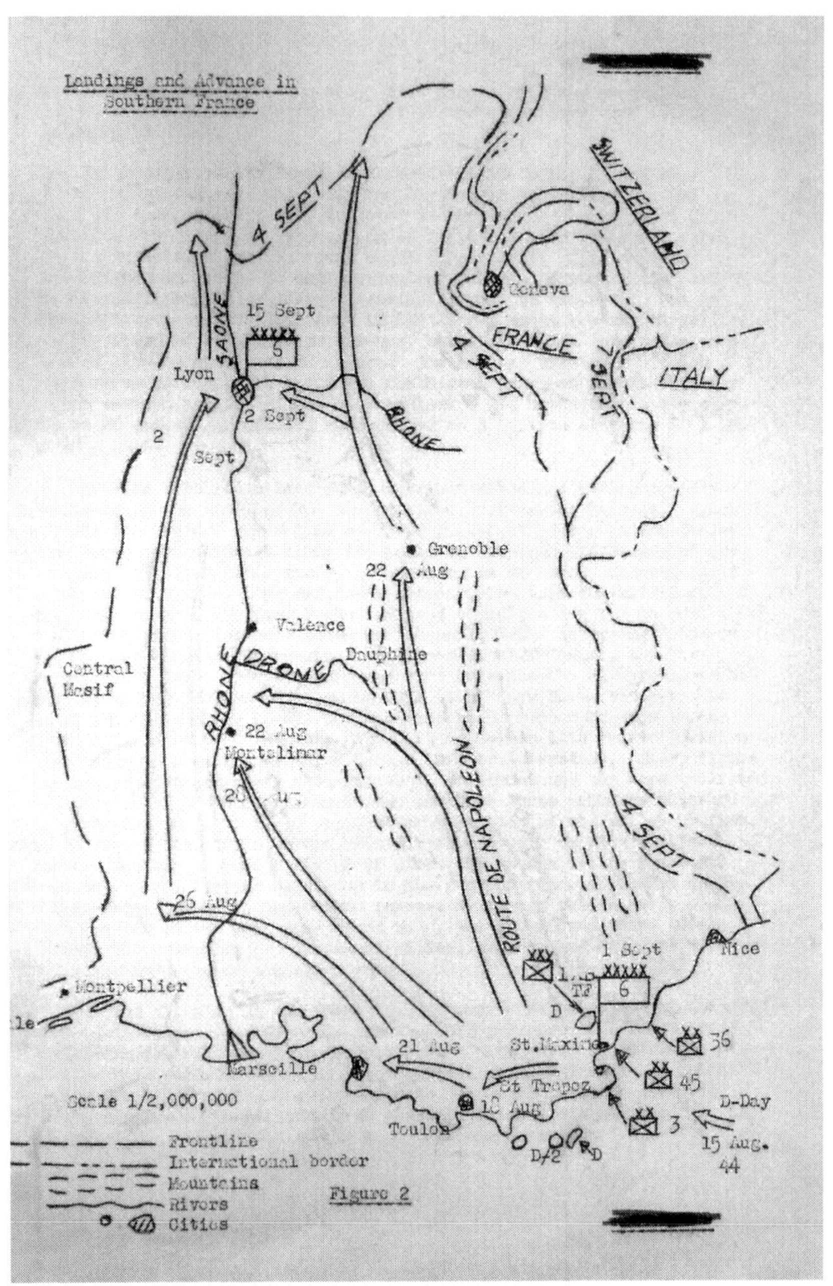

Map 6. Lines of advance up the Rhône Valley, from "Final Report: G-3 Section, Headquarters, 6th Army Group, 1945," 1945 (US Army Center of Military History, Washington, DC)

that distance. On September 15, D+30, the 6th Army Group established itself in France, and the Seventh Army had advanced to the Moselle River, a point that planners had not expected to reach until the middle of December, D+120. This quick advance was good news for commanders but presented challenges to the supporting units. Service units were still flowing into southern France, and engineers could repair or lay only so many miles of rail or pipeline per day or week. The French and American divisions were outracing their supply lines.[47]

Fortunately, the Coastal BS had begun arriving in southern France with the initial assault and worked steadily to increase its capacity to manage support operations. Operating under the control of the Seventh Army, the base section provided administrative oversight of the beaches for all supply operations from the time of the landings, greatly improving beach operations compared to the landings in North Africa and Sicily. On September 4, the SOS changed the name of the base section to the Continental BS to reflect its mission, and the 6th Army Group established a communications zone that stretched north to Moulins, Macon, Bourg, and Geneva: the rear of the Seventh Army. Southern France now held two zones—one for combat and one for support.[48]

In an effort to maintain continuity of support to the advancing force, the Continental BS decided to move its headquarters and service units forward in the area of operations. On September 18, D+18, Colonel Walter Tenny led a reconnaissance force to Grenoble, a town centered on the Allied axis of advance. However, the armies were already 120 miles past that point, and it was apparent that the Continental BS needed to extend its supply chain farther north into the Rhône Valley. Dijon had fallen on September 11 and was well suited for the base section's needs, so on September 18 the base section established an advance element in that town. Within two weeks, it transferred responsibility of operating the ports and coastal areas to the newly formed Delta BS and then, on October 1, 1944, moved its operations to Dijon.[49]

While the Continental BS was working to form a viable communications zone, Allied armies were advancing farther up the Rhône Valley. Weak German resistance encouraged commanders to push combat units as far and fast as possible in order to keep the Germans from establishing a planned defensive line. This was critical because as German units pulled north and east, they were shortening their lines of communication. In other words, the closer the front came to Germany, the less distance the German army would have to cover for shipment of supplies, whereas Allied units would have a harder time supplying themselves because of the increased distance between the front lines and the ports in southern France. To make progress, therefore, the Allies could not allow the Germans to rest.

From the start, however, shortages of fuel and trucks became an issue at all levels for the Allied units. The War Department had formed quartermaster truck companies with only one driver per truck, but units needed assistant drivers in order to operate vehicles on a twenty-four-hour basis. In addition, there was competition within the divisions for trucks to move combat forces as well as supplies. American infantry units were incapable of moving their entire formations solely with organic truck units, so additional transportation units were needed. However, along with moving the men, truck and rail units also needed to move the matériel of war: rations, ammunition, fuel, construction supplies, medical supplies, and replacement equipment. The demand for transportation exceeded the available supply of trucks and rail throughout the campaign.[50]

Although the Dragoon forces were among the most experienced in the Mediterranean and European theaters, large-scale pursuit operations were a relatively new experience for them. The previous battles in Tunisia, Sicily, and the mountains of Italy had offered glimpses of how difficult it was to support pursuit operations, but those battles involved nothing on the scale faced now in France. In southern France, the enemy was in retreat, which meant that as long as the Germans retreated, the Allies needed less ammunition but more fuel. However, planners had expected a more spirited defense and therefore had scheduled the landing of large quantities of munitions instead of fuel. Now, the theater had to quickly adjust its priorities or risk losing the initiative.

This is not to say that that the Seventh Army did not need ammunition. During the breakout of the landing areas, the 36th Infantry Division expended most of its initial munitions, causing a temporary shortage within the division. The base sections had replacement supplies in France, but the sections had to locate and then transport these supplies to the front, a process that took time and transportation resources. A 200-mile pursuit to the north followed the breakout, causing a situation wherein units needed fuel more than ammunition. Then, facing a determined German defense, the division changed its priority from fuel back to ammunition. After breaking through the resistance, the 36th then advanced another 300 miles, once again relying on timely shipments of fuel to move the division forward. Although planners fought to anticipate future demands, the enemy had a vote, which often disrupted plans. Fortunately, the lack of a German air and artillery threat aided resupply efforts, so that 90 percent of resupply operations occurred during daylight hours, allowing units to move supplies more quickly and with less risk.[51]

The speed of the Allied army's advance resulted in a doubling of the fuel requirement that had been forecast. Initially, trucks were the only means of fuel distribution because of German destruction of rail lines and bridges. However, as base section engineers repaired rail lines, the gasoline-distribution system developed into an arrangement of using short stretches of rail line between destroyed bridges and then cross-loading the fuel into trucks for movement where rail movement was no longer possible. The expanding distance between the port of Marseilles and the combat units meant that it was almost as difficult to refuel the convoy trucks hauling the fuel as it was to refuel the actual combat forces.[52]

To meet the increased demand for truck units, the Seventh Army used the vehicles from air defense units to form additional truck transport units. The Luftwaffe was not a real threat anymore, so higher headquarters could use the resources of the air defense units for support tasks. Clerks, cooks, ammunition loaders, gun sergeants, and section sergeants all abandoned their traditional duties and became truck drivers; drivers drove alone and often drove with little sleep. One such driver, Technician Fifth Grade Charlie Jones drove for four days and nights, averaging three hours of sleep a night in the cab of his truck. Jones's experience reflected the dedication of the service forces working to keep the army moving.[53]

The Allied supply lines did not begin to catch up with the forward combat units until the end of September, six weeks after the landings. A stronger German resistance at the Moselle River slowed the Allied advance, providing an opportunity for base section engineers to improve the road and rail systems leading to the North.

A Mix of Modalities

Although trucks provided the initial means of sustaining the advancing Allied armies, rail represented the most efficient means of moving men and matériel to the North. As in previous operations, the Military Railway Service provided the means to leverage the most out of the available rail networks.

Lieutenant Colonel Benjamin Decker, executive officer for the Military Rail Service, landed in southern France on D+1 and began assessing the French rail infrastructure. Military rail cars landed on the beaches on D+2, and on August 17, D+3, the first train ran from Saint-Tropez to Cogolin with eight rail cars. The distance was only 5.6 miles, but this represented the first step toward a rail system that would eventually move more than 14,000 tons per day out of Marseilles.[54]

The advance team soon found twelve locomotives and eighty rail cars northeast of Toulon and immediately placed them into service. Only 10 percent of the locomotives that had been in southern France before the war were still there, but the Allies did manage to find 200,000 of the original 400,000 freight cars. To increase capability, the Railway Service shipped diesel-electric switch engines

Table 2. Total Port Discharge and Clearance Tonnage for September through December 1944

	Total Tons Discharged	% Cleared by Road	% Cleared by Rail	% Cleared by Water
September	129,240	76	23	1.0
October	407,263	57	42	1.0
November	553,966	58	40	2.0
December	418,548	66	31	3.0

and oil-burning locomotives to France from North Africa, along with 1,000 special-purpose rail cars, such as tankers and flat cars for moving mechanized equipment.[55] There was an initial shortage of coal and fuel oil needed to operate the locomotives, but the Germans had not been able to destroy the local power plant before retreating. This allowed the use of electric locomotives along a line stretching from Chambery to Culoz, but engineers had to shore up a number of bridges before the line could come into full operation.[56]

The drive north demanded that rail lines extend as far north as quickly as possible to keep pace with the advancing Allied armies and to relieve pressure on the limited numbers of trucks available in the Southern line of Communication. As in Italy, engineers and rail workers not only had to lay new track but also had to repair bridges and other critical infrastructure destroyed by retreating German units or by Allied bombing.

A double-track rail, the Rhône Valley Railroad, served as the main means of transporting men and matériel up the Rhône Valley, and a single-track line ran through Grenoble and Dole. However, having the rail lines operational was no guarantee of resupply. Supply levels in US combat units during the first two weeks of October 1944 ran low again, in part because of the shortfall in ground transportation and in part due to the actions of French Base 901, which was holding all empty rail cars around Marseilles for the exclusive resupply of the French First Army. Unsurprisingly, as soon as the SOS became aware of the situation, officers intervened to convince the French to release their hold on the rail cars. There was little doubt of the outcome. The French held the cars, but the BSs controlled the supplies. The French agreed to release the rail cars back to general service, and the supply shortfalls for all units gradually eased.[57]

Ultimately, French rail workers and US troops rebuilt the lines that provided an important means of transporting the materials of war up the Rhône Valley. The effort included the rebuilding of forty-two bridges and more than 800 miles of track within a system of more than 4,000 miles of rail. Table 2 depicts how quickly the rail capacity developed.[58]

The 6th Army Group headquarters established itself in France on September 15 and immediately went to work organizing and directing the operations of the two combat forces as well as overseeing the communications zone. During the initial stages of Dragoon, the Seventh Army handled all logistic and administrative requirements of the force. With the establishment of the 6th Army Group, the Seventh Army could now focus on fighting the Germans in an ever-deepening zone. The beaches, ports, and responsibilities for the French First Army were now in the 6th Army Group's hands.

While trucks and rail provided the initial means of moving supplies, efforts quickly got under way as well to establish a viable pipeline network to move fuel north up the Rhône. The 697th Engineer Petroleum Distribution Company landed in southern France close to Saint-Raphaël and immediately began installing pipelines, first to fighter bases in the local area and then to the armies moving north. Fortunately, they found three French refineries that had been unused in four years but were largely undamaged and required only routine maintenance to be brought back into operation for storage of fuel. This storage capacity of more than 2 million barrels near the port provided the ability to offload tankers quickly, but the base section still needed the means to transport the fuel forward to the units.[59]

By September, engineers were constructing four-inch and six-inch pipelines north from the petroleum-storage areas of Port de Bouc. For planning purposes, one Engineer Petroleum Distribution Company had the capability of constructing 100 miles of pipeline in thirty days. Construction of the pipelines continued on a twenty-four-hour basis, seven days a week, and was a top priority. The maximum rate of advance was 10 miles per day, three times the stated capability. Within three months, the pipeline extended all the way into the Alsace region, at which point engineers temporarily halted construction because they had come within range of enemy fire. Engineers in the Continental BS eventually laid a total of 1,507 miles of pipeline in southern France, in comparison to three base sections laying only 731 miles of pipeline in North Africa.[60]

Enemy sabotage, weather, and black-market thieves had an impact on the pipeline, although in the larger picture the impact was relatively modest. Cold weather and rains during the winter slowed pipeline construction, and in at least one instance German saboteurs were able to uncouple the pipeline near Avignon, causing the loss of 2,000 barrels of fuel. Most common was the routine tapping of the pipeline by a pickax or other sharp object by black marketeers. The Continental Advance Section estimated that, overall, only 2 percent of the fuel carried by the pipeline was lost to evaporation, thieves, and sabotage. This was a small quantity in terms of percentage but a large quantity when one looks at the vast quantities of fuel that needed to be transported on a daily average.

Although pipelines required extensive resources to emplace, they proved to be an important part of the transportation triad for Dragoon's forces. Eventually, pipelines laid by army engineers carried one-third of the total supply tonnage of the force—tonnage that otherwise would have required trucks or rail tanker cars for movement.[61]

Homecoming: French First Army

The support of French forces posed special challenges for the Continental BS. Clothing presented one such challenge because many French soldiers were smaller than their American counterparts. Quartermaster units issued uniforms, shoes, and individual equipment in unit sets based on US sizing averages. A company issue of uniforms contained a set number of each uniform size. When applied to French units, this set number meant a shortage of smaller-size items and an excess of the larger-size items, requiring quartermasters to arrange for an exchange of the unusable items.[62]

In addition, the different types of French forces required various levels of support. French rearmament program units, such as in the French First Army, were authorized US logistical support. Others, such as French Forces of the Interior and resistance forces, were not authorized support from the Continental BS but needed help to continue fighting the Germans. These nonregulation forces generally received food in sufficient quantities, but clothing and equipment were more problematic.

As the Allies freed up more and more French territory, large numbers of French reservists, resistance forces, volunteers, and liberated POWs came forward to join in the fight. French authorities estimated that as many as 400,000 men and boys roamed the liberated territories of France, with half this number possessing firearms.[63]

The issue regarding development of an Allied policy toward clothing and outfitting these varied forces raged throughout September and October. General Devers lacked the resources needed to supply the 112,000 men who had already augmented the French First Army or were operating along the Atlantic coastline. The debate concerning the policy for this available manpower continued until the end of December, when the French, British, and US commands finally agreed to create 120 security battalions and eight additional French divisions by June 1945, with six of the latter to be equipped from French industry. The decision to create additional French units provided a means to provide for future French security, created additional divisions for the war effort, and gave commanders a predictable requirement for resources.[64]

As in Italy and Sicily, in France the civil populace presented another sizeable demand for resources. Civil affairs planners recognized that much of the local food, fuel, and other supplies would have been forcibly requisitioned by the German army, so three Liberty ships in every convoy from D+10 to D+40 carried food and medical supplies earmarked for French civilians. This number increased to four ships per convoy between D+41 and D+80. Once these supplies arrived in France, they were stored separately from other supplies and then provided to French authorities for distribution to the populace. This means of relying on local French governments and workers for food distribution provided several benefits. First, it allowed the base section to focus its efforts and limited transportation on supporting the combat forces. Second, it provided jobs to local Frenchmen. Third, the distribution of essential supplies helped reestablish the legitimacy of the local French governments. Last, the supplies produced goodwill among the French people and prevented any disruption in the rear areas. The benefits well outweighed the costs, and so the provisioning of the French population proved to be an essential element of the overall support effort.[65]

Approaching the Rhine

The period from October to mid-November 1944 represented the next phase of Dragoon; a period of slow fighting leading up to the Vosges Mountains. Compared to the earlier fast-paced pursuit, the fighting in the Vosges seemed more like fighting in Italy once again. Supplies were still tight. Engineers had been working to expand the rail networks into the Rhône Valley, but demand still outpaced capacity. As an example, for the period October 1–7, the Seventh Army requested rail delivery of 4,485 tons of supplies, but the 6th Army Group G4 could allocate only half of that amount—some 2,270 tons. Rail capacity was increasing, but so was demand.[66]

By the first week of October, the Seventh Army was critically short of items such as repair parts, batteries, communication wire, ammunition, and some types of weapons. The problem was not solely one of ports or distribution but instead traced back to a lack of industrial production in the United States. The worldwide demand for critical items such as trucks, ammunition, and weapons systems surpassed what the War Department had projected, and it would take time to make up the difference. For many items, relief would take three to six months. During the fall of 1944, the War Department divided all ammunition between the theaters and rationed the issuing of arms and major weapons.[67]

The advance of the 6th Army Group slowed in October and November due to a number of factors. First, the supply lines were still struggling to keep pace with forward units. Second, the hills, forests, and limited roads of the Vosges slowed movement. Last, even though German units were retreating, they still put up a fight before moving back. In the final analysis, however, the 6th Army Group noted that "logistics . . . continued to be our most formidable opponent."[68]

This operational slowdown did serve as an opportunity to build up transportation networks and to begin to stockpile supplies in the forward locations. For the first time, supply reserves began to appear near the front, and the 6th Army Group began planning for a November offensive to push its way through the mountains and then head toward the Rhine. Rather than hindering the operation, the slowdown had allowed logisticians to set the conditions for the next stage of battle.

By the first week of November, the supply situation for both the French First Army and the Seventh Army had improved to a point whereby offensive actions were once again possible. Devers, noting that the German defenses were relatively weak in the 6th Army Group sector along the Rhine, decided to take a chance and directed the Seventh Army to plan for a crossing of the Rhine before the end of December. To the north, both Bradley and Montgomery were facing stiff resistance. Patton's Third Army, just north of the Seventh Army, was stalled and unable to make any headway. Devers devised a promising but risky plan to cross the Rhine near Strasburg and then swing the Seventh Army north, where he hoped that it could destroy the German Nineteenth Army and cut off the German First Army positioned in front of Bradley's 12th Army Group. The 6th Army Group had the right enemy situation, forces, and supplies to make such an operation possible.

Enabling the strategy were the SOS units that had caught up with the advancing front lines. By the third week of November, the Seventh Army was ready to cross the river and make the first large incursion into Germany from the west. Fate and a lack of awareness in the German High Command had provided Lieutenant General Devers with an opportunity seldom found in war.

However, Eisenhower chose not to pursue this opportunity. While visiting the XV Corps Headquarters on November 24, the SHAEF commander learned about the planned river crossing and quickly ordered all planning to cease. XV Corps was instead to focus its efforts on advancing to the north to relieve the pressure on Patton's Third Army. In Eisenhower's mind, the priority of the European theater was Patton's advance into the Saar basin. Devers, hearing of the change, challenged Eisenhower on the wisdom of the decision after dinner at the 6th Army Group Headquarters later that night. Bradley, Eisenhower, and Devers argued over the theater strategy and priorities until after midnight. The debate failed to change Eisenhower's mind, and the orders to abandon a November crossing of the Rhine stood fast. That night Devers wrote in his diary, "The decision not to cross the Rhine was a blow to both Patch and myself."[69]

The decision to abandon the November Rhine crossing was confirmation of Eisenhower's broad-front strategy. The SHAEF commander was worried that a narrow incursion into Germany would prompt a German response and thus cut off or inflict heavy losses on vulnerable Allied formations. In addition, September's ill-fated Operation Market-Garden had consumed vast quantities of supplies that theater logisticians were still working to replace. The Seventh Army was the only Allied army along the Western Front to be in a position to cross the Rhine, but Eisenhower was unwilling to alter the strategy for entering Germany. Instead, the Seventh Army was to reorient its effort to the north and focus on the Vosges Mountains in an effort to help the 12th Army Group. Eisenhower wanted *all* his armies moving into Germany, not just those in the South.

One historian argues that Eisenhower's decision immobilized the 6th Army Group, delayed the crossing of the Rhine by five months, and cost 200,000 additional casualties. The Seventh Army engineer Brigadier General Garrison Davidson wrote that a crossing of the Rhine by the Seventh Army could have occurred along a two-division front, with movement north afterward to envelop the Ardennes from the rear. This strategy could possibly have preempted the Battle of the Bulge, saved some 40,000 Allied casualties, and shortened the war by a number of months. In addition, had the Allies crossed the Rhine in November, the terms of the Yalta Conference might have been different because US and British forces would have potentially been deep into Germany, thus influencing the makeup of the various national governance sectors following the war.[70]

Historians and military experts can debate exactly what might have happened had Eisenhower allowed Devers to cross the Rhine in late November 1944, but all should agree that such an advance would have forced the Germans to react. Such a reaction would most probably have had some impact on Hitler's plans for the Ardennes offensive, although the exact impact is difficult to determine. In any case, this opportunity did incur some risk but was in fact logistically supportable. Germany, meanwhile, chose to go back on the offensive.

The German counteroffensive against the Allied armies in northwestern Europe (the Battle of the Bulge or Ardennes Offensive) forced the 6th Army Group to assume the defensive in December 1944. German forces attacked through the Ardennes Forest on December 16, 1944, and quickly penetrated 50 miles past the front lines. The Germans' goal was not merely to cut the US First Army in half and capture Allied supplies but to sever the Allies' Northern Line of Communication and recapture the port of Antwerp. This strategy was devised by Hitler himself and, if successful, could have seriously delayed the Allied advance into Germany. Hitler knew that the port of Antwerp was vital to the Allied advance. By December, Antwerp was supplying 25,000 tons of matériel per day to the two Allied army groups.[71]

Eisenhower suspended any further offensive operations in southern France in order to husband resources and meet the threat to the Northern Line of Communication. The Third Army reoriented to the north. The 6th Army Group also had to reposition units to occupy the sectors left open as the Third Army moved north to deal with the German incursion, and the Allied forces in the Ardennes were the priority for all supplies and equipment.[72]

Supply officers used the pause to their advantage; this was another opportunity to resupply the army and move depots forward. The supply of K rations within the Seventh Army had dropped from twelve days of supply on November 25 to a little more than two days of supply by December 21. Army fuel levels had decreased from more than 2 million gallons to only 237,710 gallons. However, while the Seventh Army supply stocks had dropped, those in the communications zone had built up.[73] Units had simply moved beyond the reach of the supply bases. Thus, the temporary defensive period allowed the base section to shift stocks as needed to resupply units and prepare for the next phase of the operation, although fuel reserves continued to be a problem due to increased demands resulting from the cold weather.[74]

The 6th Army Group maintained what was essentially a static front from December 1944 until February 1945 as the German army established a surprisingly strong position in front of the Rhine River. Hitler personally appointed Heinrich Himmler, the Nazi SS chief, to oversee the defense of the Colmar region. However, the Battle of the Bulge was not the only German offensive in the winter of 1944–1945, just the most famous. Hitler and his generals, realizing that Eisenhower had weakened the Seventh Army in order to defeat the attack in the Ardennes, devised Operation Northwind—an attack through the Low Vosges Mountains set to start on January 1. Northwind's objective was straightforward: break through the lines of the Seventh and the French First Armies and then drive south to destroy the rest of the 6th Army Group. Hitler knew that this line of communication was a source of strength for the Allies, and he sought to eliminate it.[75]

The German First Army launched its attacks at midnight on New Year's Eve, but the 6th Army Group was well prepared. Although the US commanders had been unable to determine specific German intentions, they had known that an attack was imminent. Reports indicated a buildup of German units in the Black Forest, and intelligence officers could not locate twenty-two German Western Front divisions. Reconnaissance photos revealed new artillery emplacements. All indicators pointed to an imminent attack.[76]

Accordingly, service units had been working over the previous week to position supplies near the front lines. Additional attacks soon followed, with five German offensives occurring from January 1 to January 25. Both sides experienced a shortage of resources and poor command decisions. The German offensive effort finally ceased on January 26, when German commanders realized that their reserves were exhausted and that they were engaged in a battle of attrition—a battle they could not win.[77]

Command Modifications

The pace and efficiency of command transitions in southern France were unlike anything seen previously in the Mediterranean. The US VI Corps, headed by Major General Truscott, was initially in charge of the operation and handled the initial landings. VI Corps retained overall command for the shortest period of any of the amphibious assaults, handing responsibilities for the beaches over to the Seventh Army on August 17, D+2. This quick handover allowed Truscott to take advantage of the tactical situation and drive inland in pursuit of the German forces, a task that would have been difficult had VI Corps not been relieved of any responsibility for managing the developing beaches.

The Seventh Army took on the responsibility for the rear areas almost seamlessly and provided the right level of command to direct the reinforcing combat elements ashore as well as to develop the beaches and establish additional inland supply bases. This was also the period in which the management and control of transportation assets was critical, a mission for which VI Corps was not suited, but the Seventh Army was.

The arrival of the 6th Army Group in the middle of September came at an opportune time as the Seventh Army and the French First Army advanced farther up the Rhône Valley. The appearance of Devers in southern France relieved Patch of any communications zone responsibilities or issues of supporting the French. The Seventh Army now had to worry only about the immediate mission—getting to the Rhine.

September also represented a change in command relationships for the forces of Dragoon. Up until this time, the Seventh Army had reported to General Jumbo Wilson, the supreme Allied commander of the Mediterranean. As of the middle of September, however, the forces in southern France, now led by the 6th Army Group, reported to General Eisenhower, the European SHAEF commander. The Mediterranean theater was still responsible for providing all administrative support, but the European theater made the operational decisions, which allowed Eisenhower to direct the movements of the force in southern France without adding to the administrative burden that was overwhelming the northern communications zone. In addition, Dragoon's ports were in the Mediterranean, and much of the supplies and equipment came out of Italy. This split command relationship lasted until November 20, 1944, when the northern and southern communications zones merged.

Prioritization proved to be a primary contribution of the Army Group headquarters. The 6th Army Group supply officer established priorities for critical supplies, such as fuel, munitions, engineer equipment, and cold-weather clothing. The Transportation Division managed allocations for rail tonnage, rehabilitation of transportation routes, and truck-transfer operations. This division of responsibilities allowed General Devers to maintain visibility on the status of critical commodities and make informed allocation decisions that benefited the overall effort, not just one unit.[78] Although the work of prioritizing effort and allocating supplies and transportation resources was not very glamorous, it was hugely important. The two armies could handle the tactical fight, but the 6th Army Group was the only headquarters that could manage the allocation of resources across southern France and provide direction on prioritization.

The transition of headquarters responsibilities occurred within the support units as well. The Continental BS had begun coming ashore with the initial assault elements and worked as part of the Seventh Army. On midnight of September 8, the Continental BS assumed responsibility for the management of all beach operations from the Engineer Shore Regiment, and all supplies originating from the Mediterranean came through the Peninsular BS. Its mission complete, the Northern BS in Corsica no longer supported southern France.

Larkin directed the following changes to the support organization, effective September 26, 1944: First, the Continental BS became the Continental Advance Section, an organization that would stay close to the armies and provide all administrative support. Second, a new headquarters, the Delta BS, would manage units operating out of the ports. This new base section would handle the reception of forces, equipment, and supplies into southern France and then transport them forward. Last, the effort in southern France had grown so large that SOS NATOUSA could not manage it from Italy. To provide a more responsive system, Larkin established a new headquarters element, SOS NATOUSA Advance, to oversee the Delta BS and the Continental Advance Section.

This forward element of SOS NATOUSA positioned itself at Lyons France under the command of Brigadier General Morris Gilland, the SOS chief of staff. SOS NATOUSA Advance worked with the 6th Army Group to plan and manage support for the armies but did not conduct the actual support operations; they left these responsibilities to the Delta BS and Continental Advance Section. The new Delta BS consisted of select elements of the former Continental BS and Northern BS and was under the command of Brigadier General John Ratay, the former Northern BS commander.[79]

The mission of the Delta BS was to operate base depots near the southern French ports, maintain the required levels of theater supplies, forward supplies to the Continental Advance Section, and establish hospitals in the rear areas.

The Delta BS also took care of the communication zone facilities and handled patient evacuation, POWs, and civilian relief efforts. It was a bulky organization that had great capability but little mobility in terms of relocating the unit and its facilities.[80]

In contrast, the Continental Advance Section focused on the needs of the supported armies, requesting replenishment from the Delta BS and positioning supplies in dumps immediately behind the army area or delivering items directly to units as the situation required. The Advance Section retained a modest level of stocks so it could maintain mobility and move forward as needed. To maintain mobility and speed, the armies were allocated no more than five days' worth of rations, fuel, and ammunition. The Continental Advance Section maintained up to a fifteen-day level of supplies, but the Delta BS stored everything else. In order to limit the size of its stockpiles, the Advance Section stored only those items that were commonly in demand and had a high turnover rate. Other items with a slower turnover rate were stored by the Delta BS and were available upon request.[81]

To determine the impact of the Delta BS and Continental Advance Section, one needs to look not at the size of the headquarters elements but rather at the number of units assigned or attached to each of the respective headquarters. As of October 6, the Continental Advance Section had 86 separate companies and detachments; the Delta BS had 133. Together, these two entities supported Allied units across approximately 64,000 square miles—a territory larger than the states of Georgia and Illinois.[82]

To support the French First Army, French Base 901 also split its operations between several locations. An Advance Section of Base 901 was located at Dijon on October 5, which eventually developed into the main base, while a rear element remained at Marseilles to supply French units operating in the rear and to coordinate with the Delta BS. The separation of the two base sections (US and French) continued into mid-October, but the arrangement grew increasingly cumbersome due to the growing distances between the port and the forward element as well as due to the limited French logistic capabilities.

Base 901 suffered from a serious lack of equipment and workers. At the end of September 1944, Base 901 contained only 1,200 men and 200 vehicles. The base "was snowed under with tasks entirely out of proportion to its still meager means." The French were unable to support their own units, a fact that grew more evident as the distance between the ports and combat units grew day by day. On October 22, General de Lattre de Tassigny requested that the Continental Advance Section assume the majority of Base 901's responsibilities and provide all direct supply support to the French First Army. SOS NATOUSA agreed and subsumed part of French Base 901 into the Continental Advance Section on October 24, with Général de brigade Georges-Vincent-André Granier of the French army becoming the

Continental Advance Section deputy commander. Select elements of Base 901 that dealt with purely French matters, such as personnel, remained under French control, outside of Continental Advance Section jurisdiction.[83]

This situation illustrated just how ineffective the French logistics system had become. By asking the Advance Section to provide all supplies and support, the French were admitting that they could not take care of themselves. Ironically, this was a reversal of the situation encountered twenty-four years earlier when the AEF needed help from the French to resupply US forces along the Western Front in World War I. This time around, the US military had the capacity and resources needed to support the alliance. The French had needed material assistance ever since operations in Tunisia, but they had retained some semblance of a support capability. Such a capability was limited, however, and the demands of supporting an army moving up the Rhône Valley simply overwhelmed the inadequate French logistics system.

More change occurred in October. On October 22, Devers was relieved of his role as deputy supreme Allied commander for the Mediterranean. Up to this point of the operation, Devers had been serving in two capacities: first, as commander of the 6th Army Group and, second, as the deputy commander for General Jumbo Wilson. Lieutenant General Joseph McNarney assumed the role of deputy commander for the Mediterranean, which then allowed Devers to focus on the fighting in southern France and to report solely to Eisenhower while AFHQ could now focus solely on Italy.[84]

As mentioned earlier, Devers did not enjoy the same relationship with Eisenhower as the other three-star generals in the European theater. Eisenhower had not asked for Devers as a commander. Marshall had originally proposed Devers as one of the Normandy Army Group commanders, but Eisenhower had deferred and offered the position to Bradley instead. After that, Marshall and the War Department assigned Devers to the US theater command post in the Mediterranean. Accordingly, Devers was not part of the Normandy planning effort but ended up under Eisenhower's command due more to circumstance than to design.[85]

Other changes occurred within the 6th Army Group itself. On October 25, 1944, Major General Truscott turned command of VI Corps over to Major General Edward H. Brooks. Truscott had performed admirably in southern France and left there to take command of the Fifth Army in Italy due to the reassignment of Lieutenant General Clark to the 15th Army Group.

Southern Line of Communication

Command relationships changed further on November 20, when COMZONE MTOUSA became the Southern Line of Communication (SLOC). Larkin split the Mediterranean communications zone staff into two parts. One part remained in Italy to coordinate support for the Fifth Army, while Larkin and the rest of the staff moved to France and fell in on the SOS advance headquarters at Lyon. Tom Larkin was now free to focus on the priority effort—supporting the forces of Operation Dragoon, while reporting solely to the European SOS.[86]

The SLOC continued as a headquarters until the middle of February, when the two Allied lines of communication, one leading from the South and one from the North, linked up to form a single communications zone behind the Allied armies driving east into Germany. Major General Larkin took the SLOC staff and moved to Paris, where he became the deputy commander of the European communications zone. The SLOC Headquarters deactivated, and members of the staff integrated into the European communications zone staff. The Continental Advance Section moved forward to Nancy on February 20, 1945, where they remained until the end of the war.[87]

To assess the true impact of Operation Dragoon, one needs only look at the cumulative accomplishments of the Allied drive across France and Germany from both the North *and* the South. A single advance from either of these directions could not have achieved the same results: clearing France in six months and winning the war in Europe five months after that. Dragoon was originally supposed to have diverted enemy forces away from Normandy, but the operation instead provided a larger benefit to the effort—giving the European theater a secondary line of communication from which to pour additional divisions and supplies into France.

The assumptions included in the invasion plan for Operation Dragoon show that the Allied planners expected a firmer resistance, but the logistical plans also reflect the restraints imposed by limited amounts of shipping, assault craft, and service units. Yes, the US Seventh and French First Armies did experience problems with supplies, especially fuel, almost from the initial landings, but this shortage was caused by the rapid pace of the operation and the resulting growing distances that had to be covered to supply the advancing troops. Retreating German units were able to fall back on shortened supply lines, while every mile advanced extended the Allied lines of communication.

Some critics might fault the SOS for not providing more fuel and transportation units earlier in the landings. Certainly, one can argue, greater amounts of both would have enabled a more robust pursuit of retreating German divisions from the beaches and up through the Vosges. Such a criticism, however, fails to consider the full context of the situation facing planners in the summer of 1944.

Intelligence reports indicated that there were up to fourteen German divisions operating in southern France. Any assault force had to be able to withstand an attack from such a force or risk losing the beaches. In addition, the experiences of Sicily, Italy, and Normandy showed that Hitler had a habit of ordering his armies to resist any invasion. Planners and commanders agreed that the invasion force had to be ready to expect a stern defense; as such, the force needed supplies, such as large quantities of ammunition, to break through any resistance. Prudence required that the SOS load the vessels as it had.

What Dragoon *does* show is the remarkable level of capability, flexibly, and agility that had developed in the Mediterranean forces after two years of fighting. For the first time, the SOS had to support two field armies, one of which did not use English as a native language. In addition, the operation was a primary effort for the Mediterranean but a secondary one in the overall war effort, which placed an even greater importance on the management and prioritization of men, supplies, and equipment.

Concerning the larger effort in France, Dragoon provided everything Eisenhower had hoped. First, the southern communications zone eased the burden on the northern ports by providing alternative beaches and ports to land divisions and supplies that were unable to land in northern France due to the port congestion and the enemy situation. For example, in late September Eisenhower was able to transfer the XV Corps from Patton's Third Army to Patch's Seventh Army after Devers signaled Eisenhower on September 22 that southern France could immediately handle three additional divisions. Loaded cargo ships lying off Normandy awaiting discharge in the late summer of 1944 were redirected to Marseilles, where they were quickly offloaded. During 1944, more than 80 percent of the supplies for the 6th Army Group came through Marseilles. The ports of northern France had less capability than projected, whereas the ports of southern France came under Allied control sooner and had significantly greater capacities than projected. These advantages heightened the importance of southern France in terms of both operations and support.[88]

Four years after the war's end, General Marshall noted that the potential contributions and importance of Dragoon were not readily apparent in the summer of 1944. Many officers supported the notion of landing in southern France mostly as a means to protect Eisenhower's right flank. However, the Overlord plans failed to anticipate just how effectively the Germans could block the Normandy ports and do so for an extended period. Officers had not fully appreciated the need for the additional ports in Overlord's planning, but the need became quickly evident as the operation went on.[89]

Ever since the lost opportunities of Tunisia in December 1942, Eisenhower had realized that his role was not to fight the battles but to flow as many Allied forces and their associated support onto the European continent as possible. This realization played out in the drive across Europe. Churchill was optimistic of eventual success in the war but had only limited expectations of the size of the Allied force that could land on the continent. The prime minister hoped to have thirty-six divisions standing on the Rhine by the winter of 1944, stating, "Liberate Paris by Christmas and none of us can ask for more." With the execution of Dragoon, however, Eisenhower planned on using not only the thirty-six divisions Churchill had mentioned but also an additional fifty that could flow in from the south of France: ten from the Mediterranean and forty from the United States. The critical element of Eisenhower's strategy during the European campaign was the theater's ability to accept forces, to move them forward to the front lines, and then to get them the necessary supplies and maintenance needed to fight the battles.[90]

One officer who does not get as much recognition as he probably deserves is Lieutenant General Jacob Devers. Devers took on the role of an Army Group commander, a role for which there was little specific doctrine, and properly established and operated a theater of war in southern France. He proved to be adept at pushing but not overextending the limits of both his combat and support units. He forged two armies with different nationalities and differing expectations into a cohesive fighting force. As such, the 6th Army Group covered as much ground as it could, considering the available resources, terrain, and level of German resistance.

Of course, the operation was not perfect; few military endeavors are. Some challenges seen in previous operations repeated themselves in Dragoon. The Seventh Army did not send its ship-loading information to the navy until a very late date, but this tardiness was due to the late final decision regarding how the landings would be conducted. Priorities changed at the assault location, so ships had to divert to different beaches, and there was a shortage of labor on the beaches. Much of this trouble is attributable to the friction that is part of any major military operation. The important thing to note here, however, is that the levels of this type of disruption were less than they ever had been on any of the previous major assault operations and that the service organizations were able to adjust quickly to the changes.[91]

One of the criticisms levied against the SOS during Operation Dragoon was that the service forces stocked too much ammunition and not enough gasoline. On average, 3,000 tons of munitions landed at Marseilles every day, and stocks in base section depots eventually grew to more than 100,000 tons. This argument fails to consider the larger picture, however, and seemingly discounts the fact that once the Germans decided to make a determined defense, priorities quickly shifted from fuel to munitions. The rapid advance in August and September placed more demand on speed, distance, and fuel than on destruction of enemy units. However, when enemy resistance stiffened, as it did during October, demand for munitions greatly increased, forcing the system to shift its priorities back to ammunition. Thus, although there were few large stockpiles in any of the forward locations, stocks were available in the rear.[92]

The 6th Army Group eventually moved 900 miles inland in less than nine months. Through the combined efforts of the Army Group Headquarters, the two armies, and the SOS units operating the communications zone, they collectively overcame myriad logistic problems. The armies developed their estimated supply requirements; the Army Group furnished guidance on establishment of main supply routes and installations; and the SOS developed detailed logistic plans based on this general guidance. Without this team approach, a high level of experience, and a spirit of cooperation, the operation could not have progressed as well as it did.[93]

By the end of November, just three months after the initial landings for Operation Dragoon, southern France had experienced a complete transformation. Planners had expected a tough, slow fight up into the French mainland, but the circumstances proved different. The lack of resistance presented an opportunity to advance farther into France and toward Germany much more quickly than anticipated. To make such an adjustment required a degree of mental and organization flexibility that had not existed in the Allied force in late 1943 but did develop over time. Accordingly, the 6th Army group was able to take advantage of strategic opportunities, such as the rapid movement up to the Moselle River in the early fall of 1944.

An indirect but important contribution to the effort in France came because of the integration of members of the SLOC staff into the European communications zone staff in February 1945. Lieutenant General John C. H. Lee and the European Services of Supply/Communications Zone staff had been under severe criticism ever since the Normandy landings for a perceived mishandling of the sustainment effort. With the integration of the two staffs, the European communications zone now had a wealth of experience to draw from. A study conducted by Major General LeRoy Lutes of the Army SOS found that the European system of support relied on larger organizations and headquarters staffs than the Mediterranean system had. The service units of the Mediterranean had simply learned to conduct support operations more efficiently over the years. The COMZ commander, Lieutenant General Lee, now had a capable deputy (Tom Larkin), and the staff sections gained the experience needed to coordinate the support effort for the remainder of the war.[94]

Consider how experience made a difference between the two staffs. At the northern ports, the European communications zone instituted a system of selective unloading. It brought in a ship, unloaded only priority items, and then sent the ship back out only partially unloaded. This system speeded the unloading of some essential supplies, but it also limited the overall tonnage of supplies unloaded during a day and was extremely inefficient. This meant that the northern communications zone held ships longer than necessary and that a ship might make several trips in and out of a port before being completely emptied.

In contrast, if a ship docked in southern France, the base section offloaded the entire ship at one time, and then another ship came in. This simple procedure was the product of two years of experience. Yes, it required time to unload items that were not necessarily a priority, but the system freed up merchant ships and increased daily port tonnages, thus taking full advantage of the limited infrastructure. As such, the first act of the combined northern/southern staff was to cancel the system of selective unloading unless Eisenhower personally ordered it for a specific situation. This one change increased daily tonnage volumes at the northern ports, which translated into more efficient and effective operations.[95]

The difference between the northern and southern support staffs was notable. The Mediterranean SOS had been *conducting* support operations for the past twenty-four months, whereas the European SOS had been *planning* support of the cross-channel invasion for that long. Although some officers moved back and forth between the two headquarters, there does not appear to have been any extensive sharing of systems or procedures. Each theater operated semi-independently and according to its commander's priorities. This independence was probably due largely to the personalities of the support commanders in the European theater, especially Lieutenant General Lee, whose temperament was almost the opposite of the personable Larkin.

Another difference between the two headquarters concerned the role of technical service chiefs. Each technical service (transportation, quartermaster, and ordnance) had a senior officer assigned to each of the theaters to provide advice to the theater commander and to oversee the technical aspects of his respective operation. In the northern communications zone, the chiefs were senior officers, typically major generals, with established fiefdoms. The chiefs typically worked away from the main staff and established essentially a separate chain of command within the service forces.

Within the SLOC, the relationship was quite different. Because the Mediterranean was a secondary theater, the chiefs of the technical services were not as senior. In addition, having had to support the force from the first day of landing in North Africa, the chiefs had become more of an integral part of the AFHQ and SOS staffs. This integration produced a situation in which the technical chiefs worked with instead of against one another and the AFHQ staff, which ultimately allowed for better support of the force.[96]

The result, then, of the integration of the communications zone staffs is that the nature of the European communications zone staff changed when it wrapped the SLOC into the organization. New eyes and new experience joined Lee's headquarters, which served to improve the systems that supported the Allied pursuit over the Rhine and into Germany. The integration produced a more mature and capable headquarters to support the remainder of the war.

The integration of the two communications zone staffs, however, also experienced a number of tensions and challenges during the transition period from November 1944 to February 1945. The War Department directed that Larkin and the SLOC staff would continue to handle support for the forces of the 6th Army Group, but senior officers in Lee's headquarters insisted on bypassing Larkin and dealing directly with the Continental Advance Section and the Delta BS. In addition, there was no good system to segregate supplies as they left New York or as they landed at Marseilles, so items earmarked for a particular unit may or may not make it to the right unit. Despite the challenges, the supply systems adjusted to the new arrangements by the spring and enabled the final drive into Germany.

The integration of the two communications zones in France was a challenging effort but one that could easily be underappreciated. For all the complexity and problems involved in supplying and supporting the forces on the ground, the two support staffs worked to blend the different organizations and structures together in a manner that still supported the ongoing fight at the front while limiting disruptions. A look at the various war reports and personal accounts of division and army commanders across France indicates that most of the complaints about supply problems occurred *before* the two support organizations merged. Whether this improvement was due to a natural maturing of the Northern Line of Communication or the experience gained from the melding of the two staffs is a topic for debate.

Despite British predictions, Operation Dragoon did not cause an entire halt to the Allied advance in Italy. Throughout the last half of 1944 and into the spring of 1945, Mark Clark and the 15th Army Group continued to make advances up the Italian Peninsula, tying down or destroying divisions that otherwise could have reinforced the German defense in France and Germany. One could argue that the weakening of the 15th Army Group in Italy meant the loss of a strategic opportunity; however, this argument does not fully appreciate the difficulties the Allies would have had in exploiting any breakthrough north of the Gothic Line. Regardless, as the Allies were working to push across the Rhine, operations continued in Italy.

8

Unfinished Business

During the time I have had WACs under my command they have met every test and task assigned to them.
—General Dwight D. Eisenhower

While the US Seventh Army was busy building forces for the invasion of southern France in the summer of 1944, Lieutenant General Clark and the Fifth Army were getting ready for the drive into the north of Italy, though the Gothic Line. To support such an operation required an adjustment to the US communications zone. The Peninsular BS was still operating out of Naples, but as the army moved farther north, the line of communication lengthened, stressing transportation networks. The base section needed to move forward.

The terrain north of Rome changed into a rolling plane that supported fast-paced pursuit operations. The challenge in a pursuit is not the speed of the combat forces but the pace at which the line of communication can keep up with the advancing formations. The pursuit depended on the establishment of bridges, roads, and pipelines—all of which required tremendous effort and resources. The dependence on truck transport increased as the Fifth Army left railheads behind. Every mile north increased the distance from the port of Naples.

On June 8, 1944, the Fifth Army captured the small port town of Civitavecchia, about 50 miles north of Rome and 150 miles north of Naples. Army and navy engineers as well as a force of some 500 civilian Italians quickly went to work rehabilitating the port facilities, so that within a week the first Allied ships were discharging their cargo. The port was of moderate size, offering a maximum discharge rate of 27,000 tons per week, but it offered an alternative to Naples—a port already congested with the mounting of forces for Dragoon.[1]

By the end of June, the Fifth Army was averaging an advance of eight miles per day, and the Peninsular BS needed to keep pace with the front lines. Eight miles per day at first seems like a modest distance, but over a week it meant an additional 56 miles between the combat units and the base section. Within a month, the front lines would effectively outrun their supply base. To deal with this issue, the army captured another small port located at Piombino, about 100 miles north of Civitavecchia, on June 25. Despite heavy damage and booby traps, engineers quickly cleared the port, thus allowing discharge operations to begin on June 30. Piombino was larger than Civitavecchia, offering a weekly discharge rate of 44,009 tons per week. The two ports relieved the pressure on Naples and shortened resupply lines to the combat units. However, the Fifth Army needed even more port capacity north of Rome to support a planned breakthrough of the Gothic Line—the port of Leghorn offered such a facility.[2]

Leghorn (Livorno) is a major port along the Tyrrhenian coast approximately 300 miles north of Naples. This was an attractive alternative to Naples; Leghorn was capable of simultaneously handling eleven Liberty ships, six lighters, and one tanker. Under guidance from the Fifth Army, the 36th Infantry Division captured the port on July 19, encountering the familiar scene of damage and destruction left behind by retreating German forces. Engineers had to remove 25,000 mines from the harbor and its facilities as well as repair all of the port's cranes and other equipment. German engineers had blocked the northern port entrance with eight sunken vessels; twelve more lay across the southern entrance. US engineers worked in two shifts to blast damaged quays, fill craters, and construct berths. Fortunately, most of the city's 125,000 residents had fled the fighting, so there was less demand for civil relief supplies, and the support units could focus on opening the port. Best of all, the port was only 35 miles from the front lines.[3]

The first resupply convoy arrived at the Leghorn port on September 3, and port capacities steadily increased to a rate of 45,328 tons per week by the end of September. By the end of November, all supplies for the Fifth Army came into Leghorn, and it became the new base of operations for the Peninsular BS. The communications zone had successfully moved north and was ready to support the Gothic Line offensive.[4]

The summer of 1944 proved to be one of transition for the SOS in Italy and the Peninsular BS. The theater was busy equipping US and French forces for the upcoming invasion of southern France, while combat on the Italian Peninsula was transitioning into warfare characterized by rapid movement. For the past six months, the tactical situation in Italy had involved limited advances and heavy demands for ammunition, which had made supply planning and support a matter of routine. By November, however, the nature of the fight had changed so radically that planners could not accurately predict what the combat units might need. Brigadier General Joseph Sullivan, quartermaster for the Fifth Army, noted that during pursuit operations "the supply picture changed from day to day, hour to hour." This meant that the supporting service units had to be ready for almost any eventuality.[5]

In Italy, the demand for fuel grew exponentially as the pace of the pursuit increased. In April, the Fifth Army consumed 6,818,077 gallons of V-80 gasoline. By June, its consumption had increased to 11,947,986 gallons. There was not enough time to field additional units or new equipment—the base sections had to work with the resources on hand.[6]

The summer offensive took its toll on both men and equipment. A report from the 1st Armored Division stated that the division had been in constant combat since May 1944 and was having problems with vehicle availability. Despite an intensive maintenance program, only 15 of 172 medium tanks and 11 of 115 light tanks were serviceable, largely for lack of repair parts. All of the division's 54 ammunition-carrier half-tracks needed replacement, along with 250 cargo trucks and 350 Jeeps. The Fifth Army was tired and needed refitting before advancing farther north. During the third week of July, Lieutenant General Clark approved a pause in operations along the Arno River, which gave units the time needed to refit, reequip, and prepare for the next major push across the Gothic Line and up to the Alps.[7]

Back on the Offensive—Heading for the Po

On August 31, the Fifth Army resumed the offensive. The operational pause had accomplished its objective—the army was refreshed and able to apply tremendous firepower against the defending Germans. Allied bombers cut all of the German supply lines leading into Italy. Although not certain, the battle's outcome was favorable for the Allies.

Attacking in concert with the Fifth Army, the British Eighth Army made good progress in the East and quickly penetrated the Gothic Line by September 2 but failed to achieve a complete breakthrough. The Fifth Army also made good progress but found stiffened German resistance in the mountainous terrain surrounding Il Giogo Pass. Cleverly concealed German defensive positions made of reinforced concrete slowed the army's advance and forced the 34th Infantry Division to resort to heavy artillery bombardments. Ammunition ran low for several frontline infantry companies, forcing commanders to dig in at night, while porters carried supplies forward. A six-day battle for the pass required all of the power the Fifth Army could muster, but the plan worked. By September 18, Il Giogo Pass was in American hands, and the Germans were abandoning the Gothic Line defenses.[8]

Although the Fifth and Eighth Armies had broken through the Gothic Line, neither had been able to accomplish the primary objective of destroying the German Tenth Army and advancing north of the Po River. The Allied Combined Chiefs realized that due to the siphoning off of units for Dragoon, the 15th Army Group did not have sufficient units or resources to achieve the necessary combat ratio of a three-to-one superiority needed for an attack, so the likelihood of success in Italy was slim.

The loss of forces to Dragoon had three direct implications in Italy. First, there were insufficient forces remaining in Italy to rotate fresh troops constantly on the front lines. Second, the loss of so many service units slowed down the entire Allied advance. Third, the loss of assault craft precluded the ability to conduct amphibious landings past German defensive lines. The two operations in France were the priority, and the Allies would not divert resources away from the main effort. Italy was destined to remain a secondary effort, able to tie up German divisions but not necessarily defeat them.[9]

Modest advances with little strategic gain, other than continuing to engage the German Tenth Army, characterized the remainder of the war in Italy. The fall campaign in the Apennines led to a winter impasse. The fast pursuit of the summer regenerated to a slow fight in the mountains and the use of pack mules. Snow, ice, and fog prevented any large-scale advance. Worldwide shortages of artillery ammunition limited shipments to the Italian Peninsula. All sides used the winter to rest and regroup, hoping to return to the offensive in the spring. Clark moved up to take command of the 15th Army Group in December, while General Lucian Truscott assumed command of the Fifth Army. General Harold Alexander was promoted to field marshal and given command of all Allied forces in the Mediterranean.

The rail line that ran from Leghorn to the Po Valley and up to the Apennines operated off electricity. As with Naples, the retreating Germans destroyed all electrical-generation equipment, cut transmission lines, blew a foot-long piece out of every rail, and destroyed culverts, bridges, and tunnels. In addition, they took all locomotives and railcars with them. US Army engineers decided to rebuild the rail line north of the Arno River and used a South African Mines Brigade with the British to open tunnels. Construction materials came from local Italian mills, such as the Tierne Steel Works. Reconstruction of the line started on February 12, 1945, and finished on March 27, three days ahead of schedule. The operation involved more than 2,500 men to repair 44 miles of damaged track but provided a valuable link in the resupply network for the soon-to-be advancing army.[10]

By March, ammunition production had improved, and the theater was back to the sixty-day reserve. The final spring Allied offensive called for the capture of Bologna, an advance to the Po River, and an advance on Verona and Lake Garda. The attack began on April 9 with the Eighth Army; the Fifth Army started its attack on April 14. British and US support forces had resupplied both armies over the winter, and now the 15th Army Group possessed sufficient strength to break through the German defenses. The Fifth Army alone contained the equivalent of ten divisions: six US infantry divisions, one Brazilian infantry division, one South African armored division, one US armored division, and other Italian and US units equivalent to an additional division. Another 29,000 men were either assigned as overages or positioned in replacement pools.[11]

By April 20, both armies had breached German defenses south of the Po and were racing north into the Po River Valley. In anticipation of large numbers of surrendering enemy soldiers, the base section stored enough rations to feed 400,000 POWs for thirty days. Fighter-bombers engaged in an aerial pursuit of retreating German units. To support the fight, the Peninsular BS extended a fuel pipeline to within two miles of the front and within sight of the German force at Bologna. The force of the attack, combined with German shortages of supplies, produced a campaign that lasted only two and a half weeks.[12]

Some units did experience sporadic supply shortages, but they were isolated and quickly resolved. The 34th Infantry Division was attacking the Genghis Khan defensive line south of the Po River in mid-April and found itself stalled because of a shortage of ammunition, particularly for its mortars. After the division sent word up the chain of command that a breakthrough could not be achieved with the conservation of resources, Field Marshal Alexander lifted the ammunition restrictions, and within a few hours the 34th had its munitions and achieved the desired breakthrough.[13]

Strategically, the Allies were driving the Germans into the Alps. The 15th Army Group was pushing north from Italy, while the 6th Army Group was now moving southeast through Bavaria and toward Austria. Patton's Third Army was moving south down the Danube Valley. On April 22, the 10th Mountain Division made the initial crossing of the Po River and began closing the escape routes in the northern part of the valley. A week later, on May 2, the German surrender in Italy was complete.

Surprisingly, the intense Allied bombing effort in Italy proved to have only a negligible impact on German logistics. Despite the level of air superiority that the Allies enjoyed, they were never able to cut all of the different German lines of communication. The redundant nature of the German system allowed Kesselring's supply and transportation units to shift between transportation modes, thus ensuring a constant supply flow that could sustain defensive operations. However, the secondary nature of the theater and the other demands for German units and supplies meant that the German High Command could not give Kesselring the resources needed to break through the Allied lines.

It would be unfair to characterize the Allied campaign in Italy as a secondary effort and imply that it held little strategic value. Indeed, like the rest of the operations in the Mediterranean, the fight in Italy did not win the war, but it contributed in important ways to the overall war effort. Specifically, it forced Hitler to provide units, men, and resources that the Germans otherwise could have diverted to the Eastern and Western Fronts. In addition, occupying Italy helped clear the eastern Mediterranean of any German air or sea threat, thus allowing for secure shipments of Allied war matériel to North Africa, Italy, and southern

France. Finally, the efforts to seize, clear, and develop seaports provided valuable experience that had a direct and significant impact on later similar operations in northwestern Europe and southern France. Italy provided the training ground for establishing a communications zone, a skill lacking during Operation Torch and not necessary during Operation Husky. Italy did become a secondary effort in the Mediterranean after June 1944, but this is exactly what should have happened and does not negate the important benefits that the fight there provided to the Allied effort across the European theater: as in previous Allied operations, it matured the force and allowed the service forces to refine the systems needed to support a theater.

An Impressive Scale of Operations

Statistics provide a means of illustrating just how capable the Peninsular BS had become. The ordnance section of the base section established a storage depot plant at Leghorn starting from scratch. The plant quickly developed into a fully functional facility that could handle more than 76,000 ordnance items and consisted of 662,000 square feet of covered storage and more than 6 million square feet of open storage, the equivalent of more than forty Superdomes. One heavy-maintenance company built a vehicle-assembly plant (similar to the twin-unit pack operations of North Africa) out of a burned-out hulk of a building that had no roof. Engineers installed support trusses and a roof; mechanics installed lifts and cranes. This one plant alone assembled 5,000 light- and medium-cargo trucks per month. To provide construction materials, base section engineers operated sixty civilian mills with a monthly output of 5.5 million board feet of lumber. Outside Leghorn, quartermasters opened one of the largest depots of the war. The depot was two miles long and half a mile wide and had nine miles of internal roadways. This facility was in full operation only a month after the Fifth Army had seized the town. The Peninsular BS also contained 109 hospitals, which cared for 82,570 patients from October 1943 to April 1945, and surgical units were located closer to the front lines than ever before. A laboratory at Bagnoli, one of two of its kind in the entire war, conducted advanced research on infectious diseases and pioneered new treatments, such as whole-blood transfusions and secondary-wound closures, as well as studies on acute hepatitis.[14]

Other service units not assigned to the base sections likewise took pride in developing the best support systems possible. The quartermaster of the Fifth Army, Brigadier General Sullivan, lamented that the SOS was trying to infringe on the exceptional supply system that the Fifth Army had developed in Italy. Moreover, besides supporting US soldiers, by December 1944 Fifth Army service units were working to support up to 30,000 partisans as well. By 1945, the

US military had developed the doctrine and capability to support a theater on such a scale that Pershing would never have imagined. This capability included several segments of American society that were critical to the theater but received little attention.[15]

Not Just a Man's War

One could reasonably argue that World War II was total war: a conflict in which nations were forced to mobilize the entirety of their populations and resources to support the conflict at hand. As a result, the war served as a forcing function for the transformation of American society. Indeed, the war itself was one of transformation: transformation of doctrine, equipment, and the nature of warfare. Included in this process was a change in the makeup of American military forces.

The problem was straightforward: the growing demand for military forces of all types outpaced the nation's supply of white men of military age who were in reasonably good health. Women and people of color had served in and with the American military in previous conflicts, but in limited numbers and with differing functions, and not all of the services held the same views. During World War I, the navy and Marine Corps enlisted women to support stateside requirements, but the army refused to do so. In 1928, Major Everett S. Hughes authored a War Department study that concluded that women would play an inevitable part of the next war directly proportional to whether the conflict would approach that of total war. Hughes would later see his prediction fulfilled.[16]

Within the wartime homeland, women filled jobs previously held by men, thus releasing the men to enlist in military service. In the military services, women initially filled roles traditionally held by women, such as nursing, but then began to take on a greater variety of noncombat functions as the demand for forces increased. On May 15, 1942, the army received legislative authority to establish the Women's Army Auxiliary Corps (WAAC), which later transitioned to the Women's Army Corps (WAC) in September 1943. All services now had the ability to enlist women as a formal part of their force. By the end of the war, more than 400,000 women had served in the US military, 150,000 of them as WACs.

Women were an important presence in the Mediterranean theater from the beginning. Nurses, such as those in the 48th Surgical Hospital, landed with the assault forces of Torch in November 1942. Duty could be dangerous—in December 1942, five WAAC officers whom Eisenhower had requested to serve as executive secretaries had to be rescued by a British destroyer when their troop ship was torpedoed en route to Algiers from England. The women lost all their personal clothing and equipment, but the theater was ill prepared replace the lost items. General Marshall met these women while he was attending the Casablanca Conference. Upon return to Washington, Marshall personally paid for and shipped new clothing to the women after discovering there were no legal means to for the army to provide free replacement of the lost items. These same WACs served on both AFHQ and SHAEF staffs throughout the North African, Mediterranean, and European campaigns. Eisenhower wrote, "During the time I have had WACs under my command they have met every test and task assigned to them. . . . [T]heir contributions in efficiency, skill, spirit, and determination are immeasurable."[17]

In January 1943, the 149th WAAC Post Headquarters Company, comprising mostly bilingual telephone operations and communication specialists, arrived in Algiers to support Eisenhower's headquarters. It was the first unit to arrive overseas, and newspapers referred to it as the "first American women's expeditionary force in history." Additional postal units joined the 149th later in May. That November, a WAAC signal company arrived in North Africa to serve as radio operators, teletypists, code clerks, and radio-room tape cutters. There were not enough women in service to meet the growing demand, however; the theater adjutant general noted there was a scramble by the various staff sections to request additional WAACs.[18]

In September, the 60th WAC Headquarters arrived in North Africa to augment the 149th Post Headquarters, and the 61st WAC Headquarters arrived to serve the SOS Headquarters in Oran. Additional women arrived between November 1943 and January 1944 to support the Army Air Forces. By January 1944, there were more than 1,500 American WACs in theater, and that number would eventually grow to almost 2,000. Eventually all WACs in the Mediterranean, except those within the Fifth Army and the US Army Air Forces, were consolidated within a single battalion—the 2629th WAC Battalion located in Caserta, commanded by Major Hortense M. Boutell.[19]

Within the army, women were part of the Army Air Forces, the Army Service Forces, and the Army Ground Forces. Almost all the technical services incorporated WACs in some capacity. Women were an important part of higher-level headquarters and hospital units, found throughout the rear area and in the combat zone at army-level headquarters as well as within medical units. They brought unique skills and talents, sometimes performing duties better than their male counterparts. Every position filled by a woman released a man for combat service.

By November 1943, WAC units were beginning to move from North Africa to Italy, following the fight as it moved up the Italian Peninsula. Especially notable was the 6669th Headquarters Platoon supporting the Fifth Army Headquarters, commanded by First Lieutenant Cora Foster. This sixty-women unit was typically only 12 to 35 miles from the front, and the WACs billeted themselves in whatever makeshift facility might be available. The unit was essentially an experiment to see how women might be utilized in forward areas. The 6669th accompanied Clark's headquarters from Algeria to Naples and eventually north through the Italian mainland.[20]

Women were even receiving high-level citations for heroism. First Lieutenant Mary Ann Sullivan was awarded the Legion of Merit in Tunisia, and Second Lieutenant Anna Goforth received the same award for her work as a surgical nurse during the Sicilian campaign. A WAC private earned a Soldier's Medal for rescuing a soldier from a pool of flaming gasoline. Lieutenants Rita Rourke and Elaine Roe were given Silver Star medals for their actions to save forty-two patients during enemy shelling in Italy. Five women received Bronze Star medals for heroism at Anzio alone.[21]

Not everyone appreciated having women in a theater of operations, and there were various allegations of discipline problems among the WACs, but they were largely unfounded. In 1945, a study on the WAC in the European Theater of Operations found that men greatly surpassed women in incidence rates for eight out of nine types of offences. The incidence rate for court-martials was almost ten times greater for men than for women.[22]

A telling testament to the work performed by women in the Mediterranean comes from Major General Harold Bull, Eisenhower's operations officer both in the Mediterranean and at SHAEF. In an undated memo, Bull wrote: "In all their assignments the W.A.C.s in North Africa were performing their work seriously and with spirit and enthusiasm. Their work was of the highest quality. Their personal appearance, military stature, and soldierly conduct were outward evidence of pride and high morale. . . . [T]he W.A.C.s in North Africa have created a demand for their services through work well done."

Women were making a difference in the Mediterranean, but they were not the only disenfranchised population contributing to the war effort.[23]

Contributions of Black Americans

Like women, Black Americans were looking to contribute to the war effort as well. In 1940, Black Americans represented 9.8 percent of the US population, and participation in the war was seen not just as a patriotic duty but more importantly as a path to the realization of full citizenship. The America of the early 1940s was one of widespread racism, Jim Crow laws, segregation, and unequal educational opportunities for minorities, particularly in the American South. Black Americans demanded the right to contribute to the war effort, whether in civilian industry or in combat.

Blacks have been part of the American military since the earliest days of the nation's founding. More than 5,000 African Americans participated in the Revolutionary War. More than 186,000 Black soldiers served in the Civil War, assigned to 167 Black regiments. World War I saw the enlistment of 404,348 Black Americans, with most serving in service units. Despite this demonstrated service, questions remained deeply rooted within society and across military leadership regarding the suitability and abilities of minorities for combat.[24]

As of October 1940, the War Department had already concluded that the military would need to maintain generally the same proportion of Blacks as that found in American society, roughly 10 percent. In addition, a statement approved by President Roosevelt added that Black units would be in all the services, and Blacks would be eligible for reserve commissions, but the War Department would not integrate Blacks and whites in the same regimental organizations. Unfortunately, the utilization of Blacks and Black units would often be shaped by stereotypes that persisted in the American military, including beliefs that Blacks did not make good soldiers because they were not brave or susceptible to discipline, needed to be led by white officers, did not have the right temperament to make good sailors, and would not stand up under direct fire. The war proved these perceptions grossly inaccurate, but such misperceptions do shed light on the War Department's approach to forming Black units.[25]

During World War II, most Blacks serving in the army found themselves in combat support units. In December 1943, there were 6,778,380 enlisted men in the army, of which 754,650 (11 percent) were Black. Combat units were 5.6 percent Black. The US Army Air Force was 10 percent Black. In comparison, 28 percent of service units were Black. America was comfortable having Blacks perform "menial" tasks such as driving trucks and building roads; it was less comfortable placing Blacks in combat and in technical or high-profile jobs.[26]

As time went on and the nature of the war became more apparent, the demand for combat support units steadily increased. Troops and supplies needed to be transported across great distances. Ports took on strategic importance as crews worked first to make the facility operational and then to discharge ships at rates unheard of in prewar conditions. Engineers built roads, bridges, and airfields, often under austere conditions. Quite often, Black units were the ones performing these critical tasks.

In April 1942, 42 percent of the army's engineer units were Black. Likewise, 34 percent of all quartermaster units were Black. There were Black combat regiments, but they made up less than 5 percent of all combat units. Between 1940 and August 1942, the army would activate twenty-seven engineer general service units and then double that number again by 1945. More than 1,600 quartermaster units provided all types of transportation and field service support. Black port battalions enabled the arrival of the material of war into the theaters of war. Black transport units became a regular fixture of infantry and armor divisions. In December 1944, the army had 477,421 Black Americans deployed in overseas theaters. Of this number, 169,678 (36 percent) were in quartermaster service, 111,012 (23 percent) were in engineering, and 64,458 (14 percent) were in transportation. Black solders found themselves assigned to units that did not always receive much acclaim, but they were making a critical contribution to the war effort.[27]

The story of the 22nd Quartermaster Group is one small part of the larger Mediterranean effort but is illustrative of the contribution made by Black service units. The 22nd Quartermaster Regiment (later changed from regiment to group) landed in Casablanca on November 18, 1942, in the D+5 convoy within Patton's Western Task Force. The regiment operated convoys for American and French forces and trained 3,500 French truck drivers in 1942 using Lend-Lease vehicles. As part of the Atlantic BS, the units of the 22nd Quartermaster drove 7,311,000 truck miles, averaging 664,637 miles per month. In October 1943, the 22nd deployed to Italy as part of the Peninsular BS and began training Italian truck companies. As a group headquarters, the 22nd Quartermaster operated fifteen Black truck battalions, four white US truck battalions, two British truck battalions, and seven Italian truck battalions.

The 22nd Quartermaster Group had a strength of 13,000 by the end of the war, with 6,750 of them Black. This was about the same end strength of an American division. All three of its original battalions (37th, 110th, and 125th—all of them Black battalions) were the first truck battalions in the Mediterranean to receive the Meritorious Service Plaque. Two of the group's companies had Black commanders with white subordinates, a departure from army policy, but one that worked. The 22nd was a glimpse into a future integrated army.[28]

The 22nd Quartermaster Group and hundreds of other Black units were a critical part of the Mediterranean theater. A similar experience can also be found later in the European theater, where Black units made up much of John Lee's SOS. Perhaps the most famous support vignette is that of the Red Ball Express—the transportation effort to resupply Eisenhower's armies as they pursued retreating German forces between August and December 1944. On average, 899 vehicles were deployed each day, with the round trip averaging 53.4 hours for the run between Paris and Saint-Lô. In an eight-day period, more than 130 truck companies delivered 89,939 tons of supplies, consuming 300,000 gallons of fuel per day in the process. The story is widely told, but what is less well known is that most of the units supporting the Red Ball Express were Black. Seventy-three percent of the truck companies in the European theater consisted of Black soldiers.[29] Black Americans were still victims of inequality and racism, but they, along with WAACs/WACs, were making an undeniable difference to

Table 3. Black Units in the Mediterranean, World War II

Unit Type	Number in Mediterranean Theater
Tank Battalion	1
Cavalry	1
Chemical Warfare	3
Engineers	22
Field Artillery	5
Infantry	4
Medical	3
Military Police	9
Band	2
Signal	1
Ordnance	7
Quartermaster	89
Tank Destroyer	1
Port Battalion	2
Port Companies	8
Army Air Forces	32

the war effort. Table 3 illustrates the distribution of Black units within the Mediterranean theater.[30]

German forces surrendered in Italy effective May 2, 1945. A week later, on May 8, the German surrender in Europe was complete. Much had changed since the first landings in North Africa, but to determine a sense of the impact of operations in the Mediterranean, perhaps it is worth exploring how they influenced the planning and execution of Operation Overlord. Did the lessons of the Mediterranean carry over to the planning of the cross-channel invasion, and if so, to what degree? Had the Allies been learning the lessons of the previous nineteen months?

9

Impact and Conclusion

Instead of the needed supply, there were only promises and continual orders of what was to be done.
—Generaloberst Hans-Jürgen von Arnim, describing German theater logistics

The eventual success Allied forces experienced in Normandy and the subsequent drive across France and into Germany had its foundation in the experiences and learning that had occurred earlier in and around the Mediterranean. Indeed, the campaigns of the Mediterranean were essential to preparing the Allied militaries for the cross-channel invasion. As such, those early operations, especially the experience in planning and executing amphibious operations as well as in building and maturing a theater support base, played an indispensable role in winning the war in Europe.

Historians have offered a variety of conclusions concerning Allied Mediterranean strategy during World War II, but evidence shows that operations within the Mediterranean at the very least served as a crucible for learning, improvement, and development of specialized equipment. In addition, the maturing of the Mediterranean theater created a support base that could aid the eventual drive across Europe and serve as a means to reduce the overall operational risk within the European theater. The Allies would not have found victory in Europe as soon as they did had it not been for the development of the logistics and administrative systems, organizations, equipment, and doctrine that occurred throughout the Mediterranean campaigns. Effective administrative headquarters and support systems provided the foundation upon which victory could be built.

To evaluate the legacy of the Allied effort in the Mediterranean, one should consider the direct connections between the various Mediterranean campaigns and Operation Overlord, the invasion of northwestern Europe. Specifically, how did the campaigns of the Mediterranean change or shape doctrine? What was their impact on theater planning? Which units and commanders in Overlord had the benefit of previous experience in amphibious operations and theater support?

The Allies executed Operation Overlord in June 1944, but the planning effort for the cross-channel invasion began much earlier. The Americans had always believed that victory in Europe required a cross-channel invasion. British planners had begun thinking about a return to the European continent soon following the evacuation of the British Expeditionary Force in the summer of 1940. At the Casablanca Conference of January 1943, Allied leaders committed themselves not only to continue operations in the Mediterranean but also to begin preparations for an eventual cross-channel attack into northern France. At the Trident Conference in Washington, DC, in May 1943, the Allies set a target cross-channel invasion date of May 1944. Although Overlord was executed as a sequential operation following the invasion of Italy, the planning for it occurred in parallel with ongoing operations in the Mediterranean.

Operation Overlord

Responsibility for leading the initial planning of the invasion of France fell to the British lieutenant general Frederick Morgan, a forty-nine-year-old veteran of World War I. Morgan served briefly with the British Expeditionary Force in France in 1940 and afterward commanded I Corps of the British Home Forces, where he and his staff planned for the invasion of Sardinia. Although the invasion of Sardinia never occurred (the Allies chose to attack Sicily instead), Morgan and his staff gained valuable experience in planning an amphibious operation.[1]

In March 1943, the British appointed Morgan to serve as the chief of staff to the supreme Allied commander (COSSAC). The COSSAC staff would serve as the planning nucleus for the cross-channel attack. Morgan remained in this position until January 1944, when the COSSAC staff was absorbed into Eisenhower's SHAEF. Upon consolidation, Major General Walter Bedell Smith served as the SHAEF chief of staff and Morgan as deputy chief of staff.

Even though planning for Overlord was done in parallel with many of the Mediterranean operations, lessons were shared between the theaters. Units published after-action reports following each operation, which detailed both successes and problem areas. Commanders and staffs from COSSAC and the European SOS periodically visited the Mediterranean and participated in strategy conferences. Matériel developers back in the United States continuously worked to refine the tools of warfare. The War Department and theaters developed new training programs focused on amphibious operations and scheduled rehearsals prior to an invasion.

Many of the lessons learned from the Mediterranean are described in the US European Theater Historical Services Division study "The Administrative and Logistical History of the ETO [European Theater of Operations]" produced in 1946. The author of the study, Lieutenant Clifford Jones, was an army historian who participated in the invasion of Normandy as part of a provisional Engineer Special Brigade operating on Utah Beach. Jones based his work on a combination of official unit histories, letters, messages and cables, and firsthand observations.

As such, his study provides an excellent account of how the Mediterranean campaigns shaped the planning and execution of Overlord.[2]

One of the first impacts of Mediterranean operations on cross-channel planning occurred in late 1942 with the mounting and deployment of forces for Torch. The European SOS had the task of mounting American forces departing from Great Britain and executing a plan that was developed largely at AFHQ. The logistical support of Torch was less than efficient, often characterized by duplication of supply requests, little exchange of information across major headquarters, and confusion regarding the logistical situation of units. This low efficiency was caused in part by an immaturity and lack of experience across commands but directly led to a "more closely integrated supply and administrative system for Overlord two years later."[3]

Operation Torch was dissimilar from Overlord in that the invasion of North Africa required a long sea voyage, with amphibious forces originating from two different nations, and was limited in terms of merchant shipping due to few convoy escorts. In addition, in North Africa the Allies faced an enemy whose loyalty was questionable and whose coastal defenses were minimal. However, Torch provided significant insights into the challenges in commanding and controlling an amphibious operation, establishing beach dumps, and improving logistics throughput, both across beach and through ports.

Although successful, Torch was characterized by poor coordination, insufficient training, inadequate equipment, and poor staging of forces and equipment. The British major general J. C. Hayton, vice chief of Combined Operations Staff, noted that the problems of Torch were not due to the enemy or to unexpected surprises but "purely to lack of training and experience."[4]

As a direct response to this poor experience, the US industrial base began designing better assault craft such as LSTs and DUKWs, and the army adjusted its doctrine, removing the stipulation that it would have control over all landing craft. The mission of the engineer amphibian brigades greatly expanded to include providing logistics support until base sections could become established. The amphibian brigades became Engineer Special Brigades and tripled in size. The army and navy began cooperating on problems such as how to deal with offshore sandbars and underwater obstacles. During the winter of 1943, the Allies built an amphibious training center at Port-aux-Poules, Algeria, and a planning school was later established in the United Kingdom—both measures intended to correct the deficiencies of Torch. Perhaps most importantly, the European theater learned the necessity of unified control in all aspects of the operation. Bradley later wrote, "In Africa, we learned to crawl, to walk—then run."[5]

The invasion of Sicily, Operation Husky, was more like Overlord in that both operations involved a short sea journey followed by airborne and amphibious attacks on an enemy that was expecting an invasion. Admittedly, however, the Italian defenders of Sicily lacked the will to fight that the Allies would later face in Normandy. Husky's ineffective airborne operations illustrated how weather, inexperience, and lack of coordination could negatively affect an operation. This helps explain why the Allies were so interested in meteorological planning for Overlord and in improving the coordination of future airborne operations. As a result of the fratricide experienced in Husky, Overlord air planners routed ingress routes well away from the invasion fleet.[6]

Husky also proved a key planning assumption invalid. A basic planning assumption going into Husky was that assault beaches could support logistic operations for only a short time before beginning to deteriorate. Sicily (and later Anzio) proved, however, that beaches could support sustained logistics operations for corps-size units without significant deterioration. Indeed, with the right engineer support, even beaches determined to be of poor quality could be used effectively. This insight would allow the Allies to develop extensive beach logistics support plans for Normandy.[7]

The other major insight from Husky was the utility of the six-by-six-foot amphibious trucks known as DUKWs. Slow and less than graceful, these vehicles became the pentathletes of amphibious logistics. The DUKWs could move supplies ashore, travel beyond the beach if needed, evacuate casualties, and even transport troops. This versatility made the ungainly vehicles prone to misuse, however, and DUKWs were routinely commandeered or hijacked by commanders at all levels. Roland Ruppenthal writes that after Husky "no landing operation was to be attempted without them." Recognizing the challenges associated with maintaining these vehicles, the Army established specialized schools in the US to train DUKW operators and mechanics. In addition to temporary piers, Overlord plans called for supply ships to offload some cargo offshore and use DUKWs to ferry the material across the beach. The system worked effectively but not perfectly because the navy was risk adverse and at Normandy occasionally anchored vessels up to 15 miles offshore to avoid enemy artillery fire.[8]

Husky generated a number of other insights for European planners, including regarding the challenges in landing airborne and glider forces close to their target areas, the importance of beach groups, and the usefulness of pontoon causeways to support beach operations. Sicily also demonstrated that landing craft could survive heavy surf and high winds, thus alleviating one of the Overlord planners' greatest fears. A key operational insight was that planners should not lose sight of the operational objective. In Sicily, invasion plans stopped roughly 20 miles past the beaches. Commanders focused on ports and beaches, failing to seal the Strait of Messina. Theater planners failed to identify the strait as a key objective and allowed air planners to focus on other targets instead of insisting that heavy bombers interdict Axis naval operations between Sicily and Italy. This oversight allowed for both the enemy's reinforcement of Sicily as well as the eventual escape of Axis forces and equipment.[9]

A final impact of Husky was a change in perception among some British commanders, who had until then tended to view the American army as untrained and unreliable. Concerning the performance of the US 45th Division, the British XXX Corps commander, Lieutenant General Oliver Leese, wrote, "We were still inclined to remember the slow American progress in the early stages of Tunisia and I for one certainly did not realize the immense development in experience and technique which they had made in the last week of the North African campaign."[10]

The landings at Salerno provided COSSAC planners few logistical lessons, but they did yield a major operational insight—that German troops could be expected to take advantage of gaps between Allied forces and that these gaps represented risks to the operation. The gap between Allied forces at Salerno in September 1943 presented a weakness that Axis forces exploited and almost drove the Allies back into the sea. Learning from this, Overlord planners placed assault units in positions along a generally continuous line, with the deliberate exception of the Carentan estuary. Salerno also had a significant insight for the theater commander—the need for a commander to have every available force at his disposal. Eisenhower writes that his experience at Salerno was a key reason behind his insistence that his command of Overlord include the two Allied strategic air forces stationed in the British Isles.[11]

Operation Shingle at Anzio was the last Mediterranean assault operation to significantly shape planning for Overlord. The landings at Anzio and the subsequent challenges of moving beyond the Anzio beachhead confirmed the notion that beaches could handle long-term support operations with the right engineering. The Anzio operation also demonstrated to planners that even small ports could be useful. The port at Anzio, initially estimated at a discharge capacity of 450 to 600 tons per day by navy planners, quicky grew to a capacity of 8,000 tons by March 1944. That spring, Overlord planners modified anticipated Normandy port discharge rates based on the Anzio experience.[12]

These four operations clearly had a direct impact on cross-channel planning. Difficulties in mounting forces for Torch resulted in a new and improved system used for France as well as in new types of beach units and equipment. Husky led to improvements in planning for airborne forces. Avalanche reinforced the usefulness of DUKWs. Shingle demonstrated the value of small ports. Commanders, units, and planners of Overlord were beneficiaries of this experience, whether through direct contact, new equipment, or informed training. The historian Rick Atkinson summarizes the Allied Mediterranean experience:

"If Torch provided one benefit above all others, it was to save Washington and London from a disastrously premature landing in northern Europe."[13]

The Value of Experience

A second legacy of the Allied Mediterranean campaigns ought also to be considered—experienced leadership. From the most senior leadership down through the different levels of command, many commanders of ground, air, sea, and logistic forces gained operational experience either in the Mediterranean or from supporting Mediterranean operations.

Experience in the Mediterranean was limited at the beginning of the COSSAC planning effort because both theaters were beginning to plan for Overlord and Husky at roughly the same time. Rear Admiral Philip L. Vian, principal British staff officer for the naval branch, had a series of Mediterranean ship commands from October 1941 until March 1943. Vian spent a few months on the COSSAC staff before returning to sea duty in the summer of 1943 to participate in Operations Husky and Avalanche. He returned to England in late 1943 and ultimately commanded the naval Eastern Task Force in the Normandy landings.[14]

COSSAC underwent additional transformation in January 1944 as Eisenhower and his accompanying staff arrived to form the SHAEF. The official history of COSSAC notes that upon Eisenhower's arrival the headquarters assumed "a very real unity of aims and methods on the lines laid down for his [Eisenhower's] North African Headquarters a year earlier." The SHAEF staff and senior commanders had a great deal of Mediterranean experience. This was not coincidence, but rather a deliberate decision to use key advisers who understood the art of managing a large Allied force operating under a single unified command.[15]

The SHAEF chief of staff was Lieutenant General Walter Bedell Smith, who had been with Eisenhower since May 1942 and was a trusted confidant to and sometime hatchet man for Eisenhower. Smith had been an integral part of all the Mediterranean operations as well as a key player in the issues surrounding theater logistics and administration. He also played a significant role in identifying which AFHQ staff ought to be part of the new SHAEF.[16]

The British air chief marshal Arthur Tedder was Eisenhower's deputy supreme commander. Tedder had worked in the Mediterranean since the spring of 1942, first out of Egypt and later supporting Eisenhower as commander of the Mediterranean Air Command. Tedder's experience in North Africa, Sicily, and Italy made him well qualified to be the primary coordinator of Allied air operations in the cross-channel operation.

Several other key members of the SHAEF staff possessed Mediterranean experience as well. Major General Harold Bull served as the SHAEF operations officer, G3. Bull spent the summer of 1943 in North Africa as a special observer and joined the COSSAC staff in 1943. The British lieutenant general Sir Humfrey Gale had served with Eisenhower as the chief administrative officer at AFHQ since 1942. Eisenhower had insisted on the transfer of Gale from the Mediterranean, feeling that "he [Eisenhower] would be unwilling to undertake another large Allied Command without Gale's administrative assistance."[17]

The senior commanders for Overlord possessed a wealth of Mediterranean experience. On the ground, Field Marshal Bernard Montgomery, commanding the 21st Army Group, had commanded the British Eighth Army since August 1942, leading campaigns across North Africa, Sicily, and Italy. His American counterpart, Lieutenant General Omar Bradley, commander of the 12th Army Group, commanded II Corps in North Africa and Sicily. At the army level, Lieutenant General Miles Dempsey commanded the British Second Army in Normandy after successfully commanding the XIII Corps in North Africa, Sicily, and Italy. Lieutenant General George Patton commanded the American Third Army in France after holding corps and army commands in North Africa and Sicily. Like many commanders, Bradley took most of the experienced senior officers within his old headquarters to his new assignment.[18]

Senior allied air commanders possessed Mediterranean experience as well. Beginning in the summer of 1942, Lieutenant General Lewis H. Brereton commanded the US Army Middle East Air Force and later the Ninth Air Force. The US Army transitioned the Mediterranean Ninth Air Force into the Twelfth Air Force and then reactivated the Ninth as part of Eisenhower's European command. Brereton commanded the renewed Ninth Air Force in Europe and later commanded the First Allied Airborne Army.

Air Chief Marshal Tedder's deputy and chief of Strategic Air Forces for Europe, Lieutenant General Carl "Tooey" Spaatz, had been with Eisenhower since July 1942 and commanded the Allied Northwest African Air Force. Lieutenant General James Doolittle, famous for the Doolittle raid on the Japanese mainland in 1942, commanded the Northwest African Strategic Air Force in the Mediterranean and later the European Eighth Air Force supporting SHAEF. Air Marshal Sir Arthur Coningham commanded the Northwest African Tactical Airforce in Africa and transferred to Europe in January 1944 to help plan Overlord and command the 2nd Tactical Air Force. Major General Hoyt Vandenberg served as chief of staff within the Northwest African Strategic Air Force and later became the SHAEF deputy air commander in chief. In August 1944, Vandenberg assumed command of the Ninth Air Force.

Within the sea services, the British admiral Sir Bertram Ramsay arrived at SHAEF in May 1943 to serve as the Allied naval commander in chief. Ramsay had been the deputy naval commander for Operation Torch and had commanded the Eastern Task Force for Operation Husky. The British admiral Harold M. Burrough had several Mediterranean commands during 1942–1943 and in January 1945 succeeded Ramsay as Allied naval commander in chief when Ramsay died in a plane crash. The British rear admiral George Creasy supported the landings in North Africa while commanding the *Duke of York* and later became chief of staff to Admiral Ramsay in Europe.[19]

Perhaps most important is the change exhibited in Eisenhower between 1942 and 1944. The Eisenhower of Overlord was a very different man from that of Torch. Nineteen months of experience in a supreme command that contained three amphibious campaigns had produced an Allied commander capable of leading the Allied assault into Germany. The Eisenhower of Torch and even Husky was tentative, somewhat unsure of himself and his role. The Eisenhower of 1944 was confident and skilled in his role as theater commander: he knew his role and that of his subordinates.[20]

These leaders are illustrative of the talent the Allies called up for the assault on Europe. They, along with thousands of others possessing Mediterranean experience, represented a wealth of practical knowledge in airborne and amphibious operations—experience they would apply to the subsequent campaigns across Europe. However, experience was found not just at the top; numerous air, ground, sea, and service units benefited from insights gained in the Mediterranean as well.

A brief look at the major combat units illustrates the extent of Mediterranean influence in the European Theater of Operations. Of the twenty US divisions that landed in Normandy between June 6 and July 24, 1944, four had Mediterranean combat experience: 1st Infantry, 9th Infantry, 2nd Armored, and 82nd Airborne Divisions. Of ten UK divisions, three had served in the Mediterranean: 7th Armored, 50th Northumbrian Infantry, and 51st Welsh Infantry. In these units, combat experience ran from the commanding officer down to the lowest private. This list does not include the host of other separate brigades and battalions that served at army or corps level across the various Mediterranean campaigns.

Naval vessels experienced a similar transfer of experience between theaters. For example, two of three US battleships that took part in the cross-channel invasion had Mediterranean service. Likewise, three of four British battleships did. Of the thirty-three US destroyers that took part in Overlord, nineteen had Mediterranean tours. Of thirty-one British destroyers in Overlord, fourteen had sailed previously in the Mediterranean. The same undoubtedly goes for other types of ships, such as minesweepers, submarine hunters, and landing craft.

Within the air, it appears that the transfer of units between theaters was more limited, probably due to ongoing operations. As mentioned earlier, the

Ninth Air Force and its commander Lieutenant General Brereton had been reactivated in England in October 1943 following the initial activation in the Mediterranean. Air commands that had also served in North Africa included IX Bomber Command, IX Fighter Command, 52nd Troop Carrier Wing, and 5th Bombardment Wing. Many air units remained in the Mediterranean to support ongoing operations in Italy and southern France as well as to attack Axis targets across southern Europe.

Units outside the Mediterranean gained important experience supporting combat operations as well, experience that would be called upon for Overlord. John Lee's SOS had been operating in England since 1942 preparing for a possible cross-channel invasion when Operation Torch upended all plans and forced Lee's headquarters to come to grips with supporting an amphibious operation from thousands of miles away. Plans called for the European SOS to support the Center Task Force of Torch, but that SOS was short of personnel, untrained, and still working out systems and procedures. There were no operational plan and no prioritization from which to develop useful support plans. Witnesses described the mounting of Torch in September 1942 as a chaotic mess. Eisenhower's immediate subordinates, Walter Bedell Smith and Mark Clark, held John Lee personally responsible. In a biography of Lee, Hank Cox writes that Lee took away a very important lesson from the effort to mount and support the Center Task Force of Torch—that the supply effort must be closely tied to and synchronized with operational planning—and this lesson would serve the SOS well in 1944.[21]

Lee visited the North African theater in January 1943 to observe firsthand its administrative arrangement. While there, Lee also attended the Casablanca Conference. The discussions in North Africa with commanders such as Patton and Montgomery must have made an impression because soon after his return to England Lee was cabling Washington for additional numbers of service troops that would include port companies, quartermaster units, and engineers.[22]

Logistically, the Americans activated the Services of Supply to handle the administrative and logistical needs of US forces in the United Kingdom. Lee's views on command and control of the support effort and rear area quickly clashed with existing planning efforts. The issue of whether logistics ought to be controlled by the theater headquarters or by a theater support organization was never satisfactorily resolved within the European theater.[23]

The two theaters, Mediterranean and European, found themselves wrestling with the same problem—how to arrange command and control of theater forces. A notable difference, however, could be seen in the personalities of the senior logistics commanders. John C. H. Lee was an eccentric officer well liked by the commander of US service forces, Lieutenant General Brehon Somervell, but often disliked by frontline commanders. Tom Larkin, a previous subordinate of Lee, seems to have been universally respected by peers and senior officers alike. Both theaters wrestled with the issue of logistic command and control through 1943 and into 1944, but the Mediterranean theater managed eventually to find a workable solution loosely based on the British model of having the chief administrative officer for the theater also command the support forces. As late as May 27, only ten days prior to the invasion of Europe, confusion still reined within the European theater relating to responsibilities of theater organization on the continent.[24]

Although SHAEF was able to get Mediterranean logistics experience in having Gale as the chief administrative officer, Bedell Smith was never able to get the one officer he really wanted—Tom Larkin. In January 1944, Smith convinced Eisenhower that SHAEF should fire Lee and put Tom Larkin in place as the American chief administrative officer, but Lieutenant General Jacob Devers, the Mediterranean theater commander, refused to even consider releasing Larkin.[25]

By January 1945, the supply situation in France had become increasingly problematic. Lieutenant General Brehon Somervell, the commanding general of SOS, visited SHAEF to investigate problems that included a severe backlog in shipping, shortages of ammunition, and an ineffective personnel-replacement system. After visiting the various headquarters, Somervell recommended that Eisenhower consolidate all supply under Lee's organization and infuse officers from Larkin's SLOC into Lee's SOS staff. Devers had organized his logistics and administration system with Larkin exercising command over the execution of sustainment operations and his staff officers participating in the highest levels of planning. The proof of the concept was in the fact that Larkin's SOS had fewer support troops than warranted yet experienced none of the problems faced in the North. In short, Larkin and the Mediterranean had figured out an effective theater-sustainment system. Eisenhower agreed to the change and implemented it on January 29, 1945. The new system, with the sharing of experience from Larkin's staff, improved sustainment within the European theater, although not quite to the degree envisioned.[26]

Final Thoughts

To evaluate the logistical accomplishments of the Allies in the Mediterranean theater, one could start by assessing the reasons behind Axis failure. Field Marshal Albert Kesselring, the German commander of the Mediterranean theater, wrote that the major error of the Axis forces was the failure to fully appreciate the importance of Africa and the Mediterranean and therefore not to realize the impact that the Mediterranean could have on influencing the fate of Europe.

This underestimation of importance led to half-measures in both strategy and resourcing. Kesselring also noted that Italy failed to capture Malta, a British island in the Mediterranean, at the beginning of the war, and a lack of resources prevented the Germans from seizing the island later during the war. This failure resulted in the lack of a secure supply chain between Germany and the North African front line. As a result, the Axis supply services in North Africa could not keep pace with the speed of modern mechanized pursuit warfare.[27]

The problem started at the very top—Hitler lacked an understanding of the nature of expeditionary overseas warfare. He viewed the Mediterranean as a secondary theater and believed Axis forces there could get by on limited means. German commanders referred to the Mediterranean as a "poor man's war."[28]

General der Panzertruppe Hans-Jürgen von Arnim, the senior German commander in North Africa after Erwin Rommel's departure, attributed Axis failure in part to ambition and in part to the lack of a unified command. In pursuing their own personal strategic goals, Hitler and Mussolini expanded various fronts beyond the ability to mass and sustain their forces. In addition, Heinrich Himmler and Hermann Göring worked to field their own armies within the SS and Luftwaffe, respectively, thus adding to the demand for equipment, personnel, and supplies. All this led to competing requirements with insufficient resources. In 1951, von Arnim contrasted the Axis and Allied approaches to theater command: "Instead of taking one responsible man in the African theater and giving him all the means he required (including the Navy and Air Force) like the Allies did with Eisenhower, we had no unified plan, no defined mission, and no unified command. We had only partial solutions. . . . [N]o-one considered the details behind the battle. . . . [T]he only person who knew this information was the local commander. . . . Instead of the needed supply, there were only promises and continual orders of what was to be done." Armin continued to write on his experience in North Africa: "Supply was the worst problem. Every 14 days the Allies had an entire supply train come in. We were limited to individual streamers. . . . [B]y the time they [supplies] went through Naples and Sicily and crossed the Mediterranean, there was nothing left. . . . One had to see everything rosy at the HQs and the uncomfortable truth was not brought up."[29]

Major Richard Feige, a supply and administrative officer within the Fifth Panzer Army summarized the problem. Prior to North Africa, the German expedition into Norway was the only previous overseas deployment of the Wehrmacht. The Germans (and Italians) were slow to fully comprehend that air supremacy and sea supremacy are decisive in both combat and supply operations overseas. As such, they failed to fully understand the role that Malta played in both enabling Allied navel convoys and impeding Axis naval resupply. In modern combat, the radius of combat operations depends on supply, and supply depends on transportation. The German experience showed that combat forces depended on the supply services much greater in overseas operations, away from their strategic logistics base, than they did while operating on the European continent. This was a lesson that all combatants, Axis and Allied alike, had to learn.[30]

The Allies had the same challenges, but their approach to the Mediterranean was quite different. Frequent conferences between Roosevelt and Churchill (and including Stalin to a lesser degree) helped forge a strategy that was much more unified than that of the Axis partners, even if it was developed an operation at a time. Just as important, the Allied nations developed support units, doctrine, and training that could meet the needs of an overseas campaign. Axis forces threatened the sea lines of communication at the beginning of the war, but these threats decreased as the war went on, resulting in a more effective and predicable supply chain. In contrast to the incompatible German and Italian systems, a general commonality of equipment within Allied units helped provide higher levels of readiness and faster repair of equipment.

The SLOC in southern France proved to be the sole practical means to land the French army back on its home soil as well as to bring in additional US divisions to the European theater and support fully one-third of the force operating in Europe. The Mediterranean allowed the Allies, especially the US military, the opportunity to develop and exercise new support concepts, organizations, and equipment. Perhaps most importantly, senior commanders learned to appreciate the role that the communications zone of the theater played with regard to long-term operations. The Mediterranean was the laboratory in which the US military could learn, experiment, and adapt, all with minimal risk. The historian Douglas Porch argues that the Mediterranean served as a pivotal theater, "the critical link without which it would have been impossible for the Western Alliance to go from Dunkirk to Overlord," an assessment based in part on the assertion that the English Channel ports were incapable of supporting all three Allied Army Groups.[31]

The nature of supporting war had changed significantly since World War I. During World War II, US forces used nearly six times more supplies per man. These supplies were also bulkier. The average ton in 1918 occupied sixty-three cubic feet of space. By 1944, the average ton required ninety-nine cubic feet of space. Modern warfare needed more supplies, and many of these supplies required more space to ship.[32]

There was a greater usage of artillery in World War II. In 1918, one round of ammunition for every gun in the American Expeditionary Force equaled 102 tons. In 1944, one round per gun equaled 436 tons, a fourfold increase. Perhaps even more telling is the number of vehicles used within the theater. In 1944, the

European theater *lost* three times as much motor tonnage as the American Expeditionary Force had in total in 1918. Perhaps the best news was that medical care also drastically improved between the two conflicts. In 1918, eight men out of every hundred died of wounds. In 1944, the rate was only four soldiers per hundred. The Allied service units of World War II were handling more supplies, transporting them farther, and caring for the wounded better than ever before in the nation's history. This improvement was not happenstance and took time and effort to develop.[33]

Allied strength in the Mediterranean theater peaked in October 1944. By then, the theater had to support 918,000 US forces, 270,000 men of the Free French Army, and 21,000 men of the Brazilian Expeditionary Force—totaling more than 1.2 million men. Added to this was the need to provide supplies for the support of POWs and civilians. The theater was conducting operations in Italy and France and planning for a possible campaign in Yugoslavia. Not only was the theater supporting a large force in several locations at once, but also the overall quality of support was unequalled in history. Never before had so many men and their equipment received such good care. The Mediterranean Services of Supply and its base sections set a new standard for administrative support of a major land combat force.[34]

An important outcome of the Mediterranean campaigns was the planners' and commanders' better understanding and appreciation of the other types of units—that is, supply, communications, port, and so on—that were needed to support the combat divisions. Simply put, a division could not fight by itself; a division needed other units to help support operations. Each subsequent operation helped inform the planners as to how many support and administrative personnel needed to accompany the divisions or to follow in close support. Adhering to such factors helped shape the balanced force that was important for long-term operations. By 1944, planners and commanders from the War Department down to the divisions had grasped the notion that the soldiers assigned to a division were just the start of a more comprehensive package of forces for any operation. The term used for the combination of a division and its accompanying support forces was *divisional slice.*

The divisional slice represented the most generic of planning factors but proved very accurate for planning any large-scale operation. It had three main components: the combat division, additional corps and army-level combat forces, and support forces not within the division. The divisional slice represented the total forces needed to deploy each division. Table 4 shows the breakdown for a divisional slice in 1944 and illustrates the broad differences each nation took regarding army service forces.[35]

The table highlights several important differences among the various belligerents. First, the US and Great Britain had to include service forces in their

Table 4. Allied Division Slice by Nation, 1944

	United States	Great Britain	Germany	Russia
Average division	13,400	16,000	12,100	6,000
Average nondivision combat forces	11,300	14,000	5,000	5,840
Average nondivision service forces	18,700	10,000	0	0
Total	43,400	40,000	17,100	11,840

formations to support modern, mobile warfare in overseas lands. Operation Torch proved that a division could not fight long by itself, and the service forces provided the means to move and sustain the force in combat. However, despite the benefit, the overhead that accompanied each division was a constant source of debate among Allied leaders. Some leaders, such as Churchill, General Marshall, and Lieutenant General Lesley McNair, commander of US Ground Forces, complained throughout the war about having so many men tied up in service units. Leaders in the states worked to reduce the number of service units, while commanders in the theaters clamored for more. There was a constant tug and pull as the War Department struggled to balance the competing demands for manpower.

Germany and Russia chose a different approach to logistical support of their armies. Instead of conscripting large numbers of service forces, these nations used interior lines of communication, contractors, and civil authorities to move supplies forward from the factories. Using interior lines of communication, they could employ their own civilian rail lines to deliver supplies essentially from the factory to the front. This method worked while the battles occurred on the European mainland, but long-distance campaigns, such as in North Africa, proved the ineffectiveness of this type of system. Rommel's supply problems in Tunisia were a direct reflection of the lack of a capable German support organization. Table 4 is also a warning for those who compare sides based on the total number of divisions alone, without considering how large or capable the divisions were. The table shows that for every combat soldier, the Allied theater needed an additional two to three soldiers to serve as administrative support. As a result, any leader who focused solely on the combat forces overlooked the majority of the force needed to operate a theater. For the Americans and British, the campaigns of the Mediterranean showed that the divisions simply could not effectively fight without a substantial number of corps, army, and theater forces setting the conditions that made combat possible.

By early 1945, the European communications zone had matured to the point where supply challenges were no longer the major problem that they had been. The European theater assimilated the 6th Army Group and the SLOC, and the

Allies were driving east into Germany. There would still be occasional spot shortages of supplies until the war's end, but they were isolated and did not reflect any widespread shortage or systemic breakdown. One can debate whether this improved level of support is attributable to the growth of the lines of communication in northwestern Europe and southern France, the additional infusion of Larkin's more experienced staff into Lee's headquarters, or simply the experience that Lee's organization gained from supporting operations up to that point. Most probably, the improvement was a product of all three factors. Regardless, administrative support of the armies continued to improve, and complaints from commanders died away.

A conclusion that is not debatable is that the US military that invaded France in 1944 was very different from the force that had landed in North Africa nineteen months earlier. Had the Allies used the same equipment, beach organizations, planning factors, force balance, and load plans for a cross-channel assault as they used in Morocco and Algeria, there is a good chance the landings in France would have failed. The force simply would not have had the staying power to hold the beaches when faced with a determined German defense. Units might well have run out of supplies, and there certainly would have been insufficient port and beach units to land the additional forces and matériel needed to exploit any gains. Fortunately, the British recognized this risk and successfully kept the Americans from pursuing such a European landing early in the war. Eisenhower later wrote that the Allied effort to develop "revolutionary types of equipment was one of the greatest factors in the defeat of the plans of the German General Staff." The Mediterranean campaigns provided the time needed to develop and test this equipment.[36]

Considering the alternatives, the Allied strategy of conducting operations in the Mediterranean and then shifting the priority to the European theater proved to be the approach to end the war in Europe with the least overall risk. This indirect approach to warfare provided the time to train a conscript army and modernize the military. Most importantly, the army learned how to support ground and air forces deployed in an overseas theater. Organizations learned from their mistakes, and commanders gained an appreciation of how to assemble a balanced force—one that could take a beach, hold it, and then make a sustained drive inland. The lack of an effective communications zone in World War I was the weak link of the American Expeditionary Force. A quarter-century later, the US communications zone was the enabler of victory.

The Allies never lost a major battle in the Mediterranean or Europe due to logistics shortfalls. Yes, commanders did not always have everything they wanted to pursue their strategies, but this was more a reflection of the ongoing worldwide competition for resources and units than of anything else. Despite these periodic constraints, the standardization of equipment and level of resourcing provided to troops on the front made possible the unbroken series of victories across the campaigns. The enemy did enjoy some occasional tactical success, but it was sporadic and ultimately accomplished little other than to point out Allied deficiencies for correction. The combat forces had the task of dealing directly with the enemy, but the service units often determined the realm of the possible. The relationship was mutually inclusive: one could not exist without the other. The men and women of the Mediterranean and European base sections and Services of Supply deserve recognition for the sacrifices they made and the role they played in the victory against Germany.

The war was over in 1945, but the lessons and insights of the Mediterranean continued to live on. Five years after the Axis powers surrendered, a new conflict would break out on the Korean Peninsula. In many ways, the fight in Korea would be very similar to the challenges faced earlier in the Mediterranean: joint operations involving large numbers of naval, ground, and sea forces; coalition forces needing US support; the establishment of defensive perimeters followed by breakouts and fast-paced pursuits; amphibious landings and mountain warfare; weather even worse than that seen in Italy and the Ardennes.

The type of warfare seen in the Mediterranean and Europe during the 1940s may have since then changed in some ways, but at its essence the challenge of supporting overseas military forces in austere environments remains constant. Major military operations of today and tomorrow will almost always be joint and combined in nature. The physics of time and distance remain unchanged. Determination of requirements and force planning are still as much art as science. Coalition members often need support and outfitting. Theater commanders face myriad demands for their attention and do not have much time to focus on administrative details. Perhaps the lessons of the Allied experience in the Mediterranean will help keep future forces, as Patton said, out of trouble.

Notes

Abbreviations

CMH   US Army Center of Military History, Washington, DC
MHI   Military History Institute, Carlisle, PA
NARA   National Archives, College Park, MD

Introduction

1. Alfred Kesselring, "Concluding Remarks on the Mediterranean Campaign," Foreign Military Studies, [1946], box 62, Military History Institute (MHI), Carlisle, PA, 27, 35.

1. Operation Torch

1. Henry Eccles, *Logistics in the National Defense,* US Marine Corps Manual, NAVMC 2799, (Washington, DC: US Marine Corps, 1987), 134. By 1945, three-fourths of US Army soldiers would fill support positions.

2. Eccles, *Logistics in the National Defense,* 137.

3. Arthur Bryant, *The Turn of the Tide* (New York: Doubleday, 1957), 288.

4. Bryant, *The Turn of the Tide,* 350.

5. War Department, "Transportation Comparative Data: World War I–World War II," December 1942, MHI.

6. T. H. Vail Motter, *The Middle East Theater: The Persian Corridor and Aid to Russia* (Washington, DC: US Army Center of Military History, 1985), 3, 25; Richard Leighton and Robert Coakley, *Global Logistics and Strategy: 1940–1943 and 1943–1945* (Washington, DC: US Army Center of Military History, 1995), 347.

7. Larry Balsamo, "Germany's Armed Forces in the Second World War: Manpower, Armaments, and Supply," *History Teacher* 24, no. 3 (May 1991): 267.

8. R. L. DiNardo, *Mechanized Juggernaut or Military Anachronism: Horses and the German Army of World War II* (New York: Greenwood Press, 1991), 21; Julian Thompson, *The Lifeblood of War: Logistics in Armed Conflict* (New York: Brassey's, 1991), 53.

9. DiNardo, *Mechanized Juggernaut or Military Anachronism,* 8–9.

10. Balsamo, "Germany's Armed Forces in the Second World War," 266; DiNardo, *Mechanized Juggernaut or Military Anachronism,* 11.

11. Balsamo, "Germany's Armed Forces in the Second World War," 266.

12. Balsamo, "Germany's Armed Forces in the Second World War," 266–67.

13. DiNardo, *Mechanized Juggernaut or Military Anachronism,* 116.

14. William Frierson, "Preparations for Torch," 2 vols., US Army General Staff, Military Intelligence Division, G2, Historical Branch, US Army Center of Military History (CMH), Washington, DC, 1:7, 2.

15. Carlo D'Este, *World War II in the Mediterranean: 1942–1945* (Chapel Hill, NC: Algonquin, 1990), 8.

16. *Report by General Dwight D. Eisenhower on Operations in the Mediterranean Area 1942–1944* (London: His Majesty's Stationery Office, 1950), CMH, 7.

17. *Report by General Dwight D. Eisenhower on Operations in the Mediterranean Area 1942–1944,* 7.

18. George F. Howe, *Northwest Africa: Seizing the Initiative in the West* (Washington, DC: US Army Center of Military History, 2002), 53; Frierson, "Preparations for Torch," 1:71; *Report by General Dwight D. Eisenhower on Operations in the Mediterranean Area 1942–1944,* 7. Also see Frierson, "Preparations for Torch," 1:116, and "Report, Lessons of the Tunisian Campaign 1942–43," British Forces, British Military Training Directorate, Geog J Africa 353 Tunisia, CMH, 4.

19. "Summary of CAO Conference," August 22, 1942, subject: The Functions of This Headquarters with Particular Reference to the Size and Number of the Staff, box 55, Record Group (RG) 492, National Archives (NARA), College Park, MD.

20. Lieutenant General Robert Colglazier, interviewed by William Kearney 1984, Robert W. Colglazier Papers, MHI, 81; Dwight D. Eisenhower, *Crusade in Europe* (New York: Doubleday, 1948), 78.

21. Frierson, "Preparations for Torch," 1:15.

22. Frierson, "Preparations for Torch," 1:27.

23. *Administrative Planning* (London: British War Office, 1952), 16.

24. Frierson, "Preparations for Torch," 1:23; D. K. R. Crosswell, *Beetle: The Life of General Walter Bedell Smith* (Lexington: University Press of Kentucky, 2010), 294, 304.

25. Frierson, "Preparations for Torch," 1:23; scrapbook, inscription dated October 4, 1943, box 1, Carter Magruder Papers, MHI.

26. "Armor in the Invasion of North Africa," US Armored School, Fort Knox, KY, 1950, CMH, 35.

27. General Charles Magruder, interview by Lieutenant Colonel Charles Tucker, 1972–1973, section 2, box 1, Magruder Papers, MHI, 36. Also see "Armor in the Invasion of North Africa," 35.

28. Frierson, "Preparations for Torch," 1:38.

29. AFHQ, "Supply Plan for United States Forces in Torch," memorandum, December 4, 1942, in Frierson, "Preparations for Torch," 2:1–5.

30. James Huston, *The Sinews of War: Army Logistics 1775–1953* (Washington, DC: US Army Center of Military History, 1966), 497. Planners used the term *communications*

zone to refer to both a geographic area and a type of organization. Lowercase *communications zone* refers to the support area of a theater immediately behind the battlefield. Capitalized "Communications Zone" and its abbreviation COMZ or COMZONE are included in names of theater support organizations.

31. Second Infantry Division to Commanding General, Western Task Force, memorandum, December 26, 1942, subject: Lessons from Operation Torch, box 1, RG 338 E 50215, NARA, 10. Also see "Report of Operations, Western Task Force SOS," not dated but received at Headquarters Army SOS by December 12, 1942, Geog J Africa 353 Tunisia, CMH, 7.

32. "Western Task Force: Attack on Safi and Its Defenses," War Department Historical Division, n.d., CMH, 26; enclosure G to the report "First Conclusions Operation Torch," Army Section, Amphibious Force Atlantic Fleet, November 25, 1942, Esher C. Burkart Papers, MHI, 2.

33. "Western Task Force: Attack on Fedala and Its Defenses," War Department Historical Division, n.d., CMH, 50.

34. "Report of Lessons to Be Learned from Operation Torch," Third Infantry Division, December 26, 1942, box 1, RG 338 E 50215, NARA, 6.

35. "Report of Operations, Western Task Force SOS," 8; "Western Task Force: Attack on Fedala and Its Defenses," 9; Captain A. G. Shepard, "Report on Operation Torch," January 9, 1943, Burkart Papers, MHI, 79–80.

36. Douglas Porch, *The Path to Victory: The Mediterranean Theater in World War II* (New York: Farrar, Straus and Giroux, 2004), 349; "Western Task Force: Attack on Fedala and Its Defenses," 52.

37. "Report of Operations, Western Task Force SOS," 9, 30.

38. Quoted in H. Essame, *Patton: A Study in Command* (New York: Scribner's, 1974), 56.

39. Howe, *Northwest Africa,* 145.

40. "Western Task Force: Attack on Mehdia and the Port Lyautey Airdrome," War Department Historical Division, n.d., CMH, 48.

41. "Western Task Force: Attack on Mehdia and the Port Lyautey Airdrome," 58–59.

42. "Western Task Force: Attack on Mehdia and the Port Lyautey Airdrome," 151.

43. "Western Task Force: Attack on Fedala and Its Defenses," 52. For example, see "Report from the Western Task Force," December 30, 1942, subject: Lessons from Operation Torch, box 1, RG 338 E:50215, NARA, 7, 18–19.

44. Frierson, "Preparations for Torch," 2:appendix H, 1. Also see "Western Task Force: Attack on Mehdia and the Port Lyautey Airdrome," 152, and "Report of Operations, Western Task Force SOS," 12.

45. Brigadier General Hugh Gaffey, draft memoirs, box 3, Ernest N. Harmon Papers, MHI, 56.

46. Carlo D'Este, *Patton: A Genius for War* (New York: HarperCollins, 1995), 437; George S. Patton, diary, September 24, 1942, to March 5, 1943, George S. Patton Papers, Manuscript Division, Library of Congress (LOC), Washington, DC.

47. Patton, diary, December 23, 1942.

48. th Surgical Hospital, memorandum, December 23, 1942, subject: Lessons from Operation Torch, Geog J Africa 353 Tunisia, CMH; Second Infantry Division to Commanding General, Western Task Force, memorandum, December 26, 1942, 8.

49. Andrew McNamara, *Quartermaster Activities of II Corps thru Algeria, Tunisia, and Sicily and First Army thru Europe,* comp. Raymond McNally (Fort Lee, VA: US Army Publication, April 1955, 29.

50. "History of the 1st Engineer Amphibian Brigade HQ and HQ Company and the 531st Engineer Shore Regiment and Succeeding Units," unit history, comp. Howard G. DeVoe, 531st Engineer Association, 1983, MHI, 1–3.

51. Frierson, "Preparations for Torch," 2:appendix H, 2, 5–6.

52. McNamara, *Quartermaster Activities of II Corps thru Algeria,* 19.

53. "Western Task Force: Attack on Fedala and Its Defenses," 53.

54. Frierson, "Preparations for Torch," 2:appendix H, 2; Second Infantry Division to Commanding General, Western Task Force, memorandum, December 26, 1942, 5; "Western Task Force: Attack on Fedala and Its Defenses," 55.

55. Howe, *Northwest Africa,* 236.

56. Thompson, *The Lifeblood of War,* 56.

57. Howe, *Northwest Africa,* 54.

58. D'Este, *World War II in the Mediterranean,* 6.

59. Atlantic Base Section to Commanding General I Armored Corps, memorandum, January 22, 1943, subject: Report on Loss of Landing Boats in November 8 Attack, box 1, RG 338 E 50215, NARA.

60. D'Este, *World War II in the Mediterranean,* 10.

61. *Report by Dwight D. Eisenhower on Operations in the Mediterranean Area, 1942–1944,* 19.

62. Frierson, "Preparations for Torch," 1:127, quoting Wilson.

63. Shepard, "Report on Operation Torch," 68.

64. Quoted in Frierson, "Preparations for Torch," 1:134.

65. *History of 6th Port,* unit history booklet, Southern Line of Communication, undated, possibly 1945, MHI, 4–6.

66. Frierson, "Preparations for Torch," 1:134; Shepard, "Report on Operation Torch," 33.

67. Frierson, "Preparations for Torch," 1:139.

68. Moroccan Composite Wing, report, December 25, 1942, subject: Report of Lessons from Operation Torch, box 1, RG 338 E50215, NARA, 11.

69. *Report by General Dwight D. Eisenhower on Operations in the Mediterranean Area 1942–1944,* 18.

70. Eisenhower, *Crusade in Europe,* 118.

2. Establishing the Theater

1. William Frierson, "Preparations for Torch," 2 vols., US Army General Staff, Military Intelligence Division, G2, Historical Branch, CMH, 1:135.

2. Frierson, "Preparations for Torch," 1:130.

3. Roy Harris, "The Communications Zone Engineer," *Military Review* 24, no. 9 (December 1944): 37–38.

4. Combined Chiefs of Staff to the Combined Shipping Board, memorandum, October 22, 1942, subject: Shipment of New Vehicles, box 127, RG 492, NARA.

5. "History of the Atlantic Base Section to June 1, 1943," vol. 2, "Ordnance" section, CMH, 32.

6. "History of the Atlantic Base Section to June 1, 1943," vol. 2, "TUP-SUP" section, 31; E. A. Suttles, "The Assembly of Motor Vehicles in North Africa or How to Improvise," *Ordnance Sergeant* 6, no. 6 (December 1943): 328.

7. Suttles, "The Assembly of Motor Vehicles," 329; Lida Mayo, *The Ordnance Department: On Beachhead and Battlefront* (Washington, DC: US Army Center of Military History, 1991), 119.

8. Suttles, "The Assembly of Motor Vehicles," 330; "History of the Atlantic Base Section to June 1, 1943," vol. 2, "TUP-SUP" section, 35.

9. "American Rails in Eight Countries: The Story of the Military Railway Service, Army Transportation Corps," unit history, 1945, MHI, 7.

10. "American Rails in Eight Countries," 5.

11. "History of the Atlantic Base Section to June 1, 1943," vol. 2, "Planning" section, 13.

12. "American Rails in Eight Countries," 7.

13. D. K. R. Crosswell, *Beetle: The Life of General Walter Bedell Smith* (Lexington: University Press of Kentucky, 2010), 366.

14. "History, Communications Zone, NATOUSA," undated, possibly 1945, declassified in 1948, part 1, CMH, 3.

15. "Gale, Lt Gen Sir Humfrey Middleton (1890–1971)," summary of collection, King's College London, Archive Catalogues, at https://kingscollections.org/catalogues/lhcma/collection/g/ga40–001.

16. "History of Allied Force Headquarters (AFHQ)," part 2, sec. 1, December 1942–December 1943, CMH, 168; AFHQ General Order Number 13, November 2, 1942, and AFHQ General Order Number 38, December 30, 1942, box 1533, RG 492, NARA. Also see "History of Allied Force Headquarters," part 2, sec. 1, 169, and "History, Communications Zone, NATOUSA," part 1, 3.

17. AFHQ General Order Number 38.

18. AFHQ General Order Number 38. Also see "History of Allied Force Headquarters," part 2, sec. 1, 170.

19. Dwight D. Eisenhower, *Crusade in Europe* (New York: Doubleday, 1948), 137.

20. "History of Allied Force Headquarters," part 2, sec. 1, 173.

21. "History of the Atlantic Base Section to June 1, 1943," vol. 2, "Ammunition" section, 18–19.

22. "History of the Atlantic Base Section to June 1, 1943, vol. 2, "The Plan" section, 11.

23. "History of the Atlantic Base Section to June 1, 1943," vol. 2, "Ammunition" section, 19.

24. Rick Atkinson, *An Army at Dawn: The War in North Africa 1942–1943* (New York: Holt, 2002), 414.

25. Christopher Rein, *The North African Air Campaign: U. S. Army Air Forces from El Alamein to Salerno* (Lawrence: University Press of Kansas, 2012), 101; Carlo D'Este, *World War II in the Mediterranean: 1942–1945* (Chapel Hill, NC: Algonquin, 1990), 13; Crosswell, *Beetle,* 378.

26. Both Patton's and Hughes's diaries are at the Library of Congress.

27. Crosswell, *Beetle,* 436; "History of Allied Force Headquarters," part 2, sec. 1, 193.

28. Eisenhower to Hughes, memorandum, February 9, 1943, subject: Instructions, RG 492, box 124, NARA.

29. NATOUSA General Order Number 4, February 12, 1943, in "History of Allied Force Headquarters," part 2, sec. 1, 196; Hughes to Larkin, memorandum, March 27, 1943, subject: Civil Goods, box 2730, RG 492, NARA. Also see "History of Allied Force Headquarters," part 2, sec. 1, 196.

30. AFHQ General Orders Number 19, February 14, 1943, and Number 38, June 7, 1943, box 1533, RG 492, NARA.

31. General Order Number 5, subject: Eastern Base Section, February 13, 1943, box 1531, RG 492, NARA. Also see "History of the Eastern Base Section 22 February to 1 June 1943," 7 vols., CMH, 1:6.

32. Lieutenant Robert Colglazier, interviewed by William Kearney, 1984, Robert W. Colglazier Papers, MHI, 88, 90.

33. R. Manning Ancell, with Christine M. Miller, *The Biographical Dictionary of World War II Generals and Flag Officers* (Westport, CT: Greenwood Press, 1996), 348, s.v. "Arthur R. Wilson"; Hughes to Commander in Chief, AFHQ memorandum, February 6, 1943, subject: Implications of Attached Recommendations, box 127, RG 492, NARA.

34. George S. Patton, diary, September 24, 1942, to March 5, 1943, George S. Patton Papers, Manuscript Division, Library of Congress, Washington, DC.

35. Ancell, *The Biographical Dictionary,* 182, s.v. "Thomas Larkin"; "History, Communications Zone, NATOUSA," part 1, 1.

36. Quoted in "History of the Eastern Base Section," 9.

37. "History, Mediterranean Base Section, September 1942–May 1944," CMH, 19.

38. "History of Allied Force Headquarters," part 2, sec. 1, 200.

39. "History of Allied Force Headquarters," part 2, sec. 1, 201; "History, Communications Zone, NATOUSA," part 3, 5; "History of Allied Force Headquarters," part 2, sec. 1, 202.

40. Wendell Little, "Functions and Operations of G4 Division, NATOUSA," *Military Review* 24, no. 11 (February 1945): 55.

41. Quoted in Rein, *The North African Air Campaign,* 107.

42. Colglazier, interview, 94, 90.

43. "History, Communications Zone, NATOUSA," part 1, 30.

44. Hughes to Larkin, May 31, 1943, box 2730, RG 492, NARA.

45. "History, Communications Zone, NATOUSA," part 1, 35, 41. By then, the SOS had 687 service units authorized, but only 348 on hand. See General Larkin to General Pence, memorandum, May 30, 1943, no subject, box 2730, RG 490, NARA.

46. Colonel Harold Bartron to Commanding General, Eastern BS, memorandum, June 13, 1943, subject: III Air Service Area Command Areas and Dumps, box 3020, RG 492, NARA.

47. *Report by General Dwight D. Eisenhower on Operations in the Mediterranean Area 1942–1944* (London: His Majesty's Stationery Office, 1950), CMH, 25.

48. "History of Allied Force Headquarters," vol. 2, part 1 [*sic*], 105.

49. Marcel Vigneras, *Rearming the French* (Washington, DC: US Army Center of Military History, 1989), 9.

50. Chief of Staff, US Army, for Combined Chiefs of Staff, memorandum, November 11, 1942, subject: Equipment for Eight French Divisions in North Africa, box 127, RG 492, NARA; Vigneras, *Rearming the French,* 17–18.

51. "History of Allied Force Headquarters," vol. 2, part 1 [*sic*], 107–8; Vigneras, *Rearming the French,* 108.

52. Vigneras, *Rearming the French,* 35.

53. "History of the Atlantic Base Section to June 1, 1943," vol. 2, 43.

54. Vigneras, *Rearming the French,* 402. Eisenhower faced a complex and volatile situation with the Free French forces, which significantly complicated his job as a theater commander. French lines of command and control were unclear: Admiral Darlin was assassinated on December 24, 1943, and General De Gaulle was working to be recognized as the leader of all Free French.

55. General Hughes to General Wilson, April 16, 1943, box 2730, RG 492, NARA; "Ordnance in the North African and Mediterranean Theater of Operations as Recalled by Colonel Connerat," box 3, Sidney Matthews Papers, MHI, 4.

56. George F. Howe, *Northwest Africa: Seizing the Initiative in the West* (Washington, DC: US Army Center of Military History, 2002), 292.

57. Howe, *Northwest Africa,* 496.

58. Erwin Rommel, *The Rommel Papers,* ed. B. H. Liddel Hart (New York: Da Capo, 1988), 287.

59. Hans-Henning von Holtzendorff, Generalmajor, "Reasons for Rommel's Success in Africa 1941/1942," US Historical Division MS D-024, 1947, MHI, 4–5.

60. Conrad Seibt, Generalmajor, "Railroad and Air Transport for Supply of Africa through Italy" US Historical Division MS D-093, 1947, 1; Alfred Toppe, Majorgeneral, "Desert Warfare," US Historical Division MS P-129, 1954, 12; Plagemann, General Intendant (Administrative Services), "Administrative Supply of the Luftwaffe in Africa and Italy," US Historical Division MS D-002, 1947, 5.

61. Rommel, *The Rommel Papers,* 306.

62. Rommel, *The Rommel Papers,* 345, 364.

63. Alfred Kesselring, "Concluding Remarks on the Mediterranean Campaign," US Historical Division MS C-014, 1954, 27, 35.

64. Albert Kesselring, "Questions Regarding the General Strategy during the Italian Campaign," US Historical Division MS B-270, 1954, 32.

65. Martin van Creveld, *Supplying War* (Cambridge: Cambridge University Press, 1977), 199.

66. Creveld, *Supplying War,* 195, 200.

67. Paul Deichmann, General der Luftwaffe, "Mission of OB Sued with the Auxiliary Battle Command in North Africa after the Allied Landings," US Historical Division MS D-067, 1954, 53.

3. The Fall of Tunisia

1. George F. Howe, *Northwest Africa: Seizing the Initiative in the West* (Washington, DC: US Army Center of Military History, 2002), 347.

2. Howe, *Northwest Africa,* 371; Carlo D'Este, *World War II in the Mediterranean: 1942–1945* (Chapel Hill, NC: Algonquin, 1990), 15.

3. Leo Meyer, "The Strategic and Logistical History of MTO," undated manuscript, CMH, ix–7. Also see Howe, *Northwest Africa,* 390.

4. Douglas Porch, *The Path to Victory: The Mediterranean Theater in World War II* (New York: Farrar, Straus and Giroux, 2004), 385.

5. Meyer, "Strategic and Logistical History of MTO," ix–20; Andrew McNamara, *Quartermaster Activities of II Corps thru Algeria, Tunisia, and Sicily and First Army thru Europe,* comp. Raymond McNally (Fort Lee, VA: US Army Publication, April 1955), 54.

6. McNamara, *Quartermaster Activities of II Corps,* 55.

7. Porch, *The Path to Victory,* 389. Also see McNamara, *Quartermaster Activities of II Corps,* 53–60; Erwin Rommel, *The Rommel Papers,* ed. B. H. Liddel Hart (New York: Da Capo, 1988), 407.

8. Meyer, "Strategic and Logistical History of MTO," ix–24; Rick Atkinson, *An Army at Dawn: The War in North Africa, 1942–1943* (New York: Holt, 2002), 414.

9. Rommel, *The Rommel Papers,* 410.

10. Conrad Seibt, Generalmajor, "Railroad and Air Transport for Supply of Africa through Italy," US Historical Division MS D-093, 1947, 3–6.

11. Fredendall was relieved on March 5 for what was viewed as ineffective leadership. As a corps commander, he appears to have had little impact on the American logistics situation, however. See H. Essame, *Patton: A Study in Command* (New York: Scribner's, 1974), 72.

12. "History of the Eastern Base Section 22 February to 1 June 1943," vol. 1, CMH, 13, 20.

13. Howe, *Northwest Africa,* 604; "History of the Eastern Base Section," 1:48.

14. Pence to Larkin, message, May 18, 1943, box 3020, RG 492, NARA.

15. Minutes of supply meeting, NATOUSA, February 27, 1943, box 108, RG 492, NARA; "Report of Colonel James Edmunds on Supply in NATO," in "Report of Observers, Mediterranean Theater of Operations, US Army Ground Forces Observer Board," vol. 2, US Army Ground Forces Observer Board, 1942–1945, June 26, 1943, MHI, 1–2.

16. Porch, *The Path to Victory,* 406–7.

17. *Report by General Dwight D. Eisenhower on Operations in the Mediterranean Area 1942–1944* (London: His Majesty's Stationery Office, 1950), CMH, 41; "History of the Eastern Base Section," 1:46.

18. Meyer, "The Strategic and Logistical History of MTO," ix–29.

19. "History of the Eastern Base Section," 1:25–26.

20. Lieutenant General Robert Colglazier, interviewed by William Kearney, 1984, Robert W. Colglazier Papers, MHI, 103–4.

21. "History of the Eastern Base Section," 1:42.

22. "History of the Eastern Base Section," 1:43.

23. Porch, *The Path to Victory*, 409. Also see Howe, *Northwest Africa*, 609–10.

24. Howe, *Northwest Africa*, 613.

25. Howe, *Northwest Africa*, 619; Meyer, "Strategic and Logistical History of MTO," x–20.

26. Howe, *Northwest Africa*, 648.

27. D'Este, *World War II in the Mediterranean*, 36.

28. Howe, *Northwest Africa*, 666. There is no detail specifying what percentage was German versus what percentage was Italian. See McNamara, *Quartermaster Activities of II Corps*, 71, and Howe, *Northwest Africa*, 666.

29. "History, Communications Zone, NATOUSA," part 1, undated, probably 1945, declassified in 1948, CMH, 11, 26.

30. Howe, *Northwest Africa*, 680; Porch, *The Path to Victory*, 413; Christopher Rein, *The North African Air Campaign: U. S. Army Air Forces from El Alamein to Salerno* (Lawrence: University Press of Kansas, 2012), 128.

31. D'Este, *World War II in the Mediterranean*, 17.

32. Hans-Jürgen von Arnim, "Tunisian Campaign," US Historical Division MS C-094, 1951, 7–8.

33. Major Richard Feige, "The Relationship between Operations and Supply in Africa" US Historical Division MS D-125, 1947, 6–8.

34. D'Este, *World War II in the Mediterranean*, 23.

4. Operation Husky

1. Ronald Spector, *Eagle against the Sun: The American War with Japan* (New York: Free Press, 1985), 187.

2. Winston Churchill, *The Hinge of Fate* (Boston: Houghton Mifflin, 1950), 692–93.

3. Robert Dallek, *Franklin D. Roosevelt and American Foreign Policy* (New York: Oxford University Press, 1995), 367; Douglas Porch, *The Path to Victory: The Mediterranean Theater in World War II* (New York: Farrar, Straus and Giroux, 2004), 415.

4. *Report by General Dwight D. Eisenhower on Operations in the Mediterranean Area, 1942–1944* (London: His Majesty's Stationery Office, 1950), CMH, 71; Carlo D'Este, *World War II in the Mediterranean: 1942–1945* (Chapel Hill, NC: Algonquin, 1990), 48.

5. "Western Naval Task Force, the Sicilian Campaign," action report, box 4, Kent Hewitt Collection, US Navy Historical Center, Navy Yard, Washington, DC, 17.

6. Quoted in D'Este, *World War II in the Mediterranean*, 51.

7. Dwight D. Eisenhower, *Crusade in Europe* (New York: Doubleday, 1948), 163–64. Also see Field Marshal Alexander, interviewed by George Howe, n.d., comp. Sidney Matthews, box 2, Sidney Matthews Papers, MHI, 10.

8. Directives, HQ Force 141, April 7, 1943, box 124, RG 492, NARA. Patton was transferred from command of II Corps on April 15 to allow him to focus on Operation Husky.

9. Bernard Stambler, "Campaign in Sicily, Part I," manuscript, US Army Historical Division, n.d. [1945–1946?], CMH, 57. Also see Albert Garland and Howard Smyth, *Sicily and the Surrender of Italy* (Washington, DC: US Army Center of Military History, 2002), 60–63.

10. Garland and Smyth, *Sicily and the Surrender of Italy*, 81

11. Garland and Smyth, *Sicily and the Surrender of Italy*, 54; *Report by General Dwight D. Eisenhower on Operations in the Mediterranean Area, 1942–1944* (London: His Majesty's Stationery Office, 1950), CMH, 79. Allied planners assumed that any port would sustain damage from retreating Axis forces, but the exact magnitude would not be known until the port was captured. Engineer units such as the 540th Engineer Shore Regiment were designed to rehabilitate ports as well as clear beaches.

12. *Report by General Dwight D. Eisenhower on Operations in the Mediterranean Area, 1942–1944*, 80.

13. Muller to Stanley, Seventh Army memorandum, May 23, 1943, no subject, box 2737, RG 492, NARA. Also see "Western Naval Task Force, the Sicilian Campaign," 67.

14. "Report on Operation Husky by Army Section of Commander, Amphibious Force, U. S. Atlantic Fleet and Army Observers, Amphibious Force, U. S. Atlantic Fleet," June 6, 1943, MHI, 3.

15. *Administrative Planning* (London: British War Office, 1952), 25, emphasis added.

16. "Supplement to Administrative Instructions Number 10," SOS NATOUSA, June 25, 1943, box 2737, RG 492, NARA; "History of Ordnance Service in the Mediterranean Theater of Operations, November 1942 to November 1945," vol. 1, CMH, 49.

17. "Notes on the Planning and Assault Phase of the Sicilian Campaign," written by an anonymous British officer, *COHQ Bulletin*, no. Y/1 (October 1943), box 3021, RG 492, NARA, 4.

18. "Supplement to Administrative Instructions Number 10," 3.

19. "Notes on the Planning and Assault Phase of the Sicilian Campaign," 6.

20. "Report on Operation Husky by Army Section of Commander, Amphibious Force," 2.

21. Eastern Base Section to SOS NATOUSA, memorandum, September 20, 1943, subject: Administrative Lessons from Operations in Sicily, box 2775, RG 492, NARA; "EBS Planning for Operation Husky," report, n.d., box 3021, RG 492, NARA. "Mounting" refers to the staging and loading of personnel, equipment, and supplies at a port of embarkation.

22. "EBS Planning for Operation Husky," 2.

23. "EBS Planning for Operation Husky," 2; Rick Atkinson, *The Day of Battle: The War in Sicily and Italy, 1943–1944* (New York: Holt, 2007), 33–34.

24. Record of Meeting, May 20, 1943, subject: Planning for the Mounting of Operation Husky, box 2704, RG 492, NARA, 3; "Western Naval Task Force, the Sicilian Campaign," 17.

25. Record of Meeting, subject: Planning for the Mounting of Operation Husky, 6.

26. Record of Meeting, subject: Planning for the Mounting of Operation Husky, 6.

27. "Minutes of Conference Held at HQ Tunisia District," June 15, 1943, subject: Mounting of Husky in Tunisia, box 3020, RG 492, NARA, 3.

28. "Minutes of Conference Held at HQ Tunisia District," June 15, 1943, 1.

29. "Report of Force Headquarters Section (Army), Amphibious Force, Forward Echelon, 9 Aug 43," in "Report on Operation Husky by Army Section of Commander, Amphibious Force, U. S. Atlantic Fleet and Army Observers, Amphibious Force, U. S. Atlantic Fleet," CMH, 17.

30. Carlo D'Este, *Bitter Victory: The Battle for Sicily, 1943* (New York: Dutton, 1988), 156.

31. "Report of Operations, 1st Embarkation Group," May–August 1943, box 3012, RG 492, NARA, 9; D'Este, *Bitter Victory,* 157; Christopher Rein, *The North African Air Campaign: U. S. Army Air Forces from El Alamein to Salerno* (Lawrence: University Press of Kansas, 2012), 146.

32. H. H. Dunham. "US Army and the Conquest of Sicily," monograph prepared by the Army Transportation Corps, 1945, CMH, 1.

33. D'Este, *Bitter Victory,* 199.

34. D'Este, *Bitter Victory,* 229.

35. D'Este, *Bitter Victory,* 231. Also see Garland and Smyth, *Sicily and the Surrender of Italy,* 121.

36. Garland and Smyth, *Sicily and the Surrender of Italy,* 117.

37. Garland and Smyth, *Sicily and the Surrender of Italy,* 119.

38. D'Este, *Bitter Victory,* 257.

39. "Report of Operations: 'Shore Engineers in Sicily,'" August 11, 1943, box 1, RG 338 E:30215, NARA, 2. Also see Alfred Beck, Abe Bortz, Charles Lynch, Lida Mayo, and Ralph Weld, *The Corps of Engineers: The War against Germany* (Washington, DC: US Army Center of Military History, 1998), 122.

40. "AFHQ Administrative Lessons, Operations in Sicily," October 1943, box 2775, RG 492, NARA, 26.

41. "Report of Force Headquarters Section (Army), Amphibious Force, Forward Echelon, 9 Aug 43," 4; "AFHQ Administrative Lessons, Operations in Sicily," 30.

42. "First Partial Report, Observations—Husky—Joss Task Force," July 8 to July 12, 1943, box 2775, RG 492, NARA, 1. Also see "Notes on Husky Landings," July 23, 1943, box 3021, RG 492, NARA, 2; "Western Naval Task Force, the Sicilian Campaign," 54; Atkinson, *The Day of Battle,* 80, 87.

43. E. S. Van Duesen, "Trucks That go Down to the Sea," *Army Ordnance Magazine* 25, no. 141 (November–December 1943): 558. "DUKW" was not an acronym but rather part of the industry naming nomenclature, which indicated the model, year, type of drive, and equipment. Each vehicle could carry two and a half tons and travel at speeds of up to 50 miles per hour on roads or 5.5 knots over calm water.

44. "First Partial Report, Observations–Husky–Joss Task Force," 8.

45. "Notes on the Planning and Assault Phase of the Sicilian Campaign," 22.

46. "First Partial Report, Observations—Husky—Joss Task Force," 12.

47. "First Partial Report, Observations—Husky—Joss Task Force," 12, 20.

48. "First Partial Report, Observations—Husky—Joss Task Force," 13.

49. "Report of Operations: 'Shore Engineers in Sicily,'" 5.

50. Duesen, "Trucks That Go Down to the Sea," 558.

51. "Report of Operations: 'Shore Engineers in Sicily,'" 1.

52. AFHQ memorandum, July 23, 1943, subject: Notes on Husky Landings, box 492–3021, NARA, 1; "Western Naval Task Force, the Sicilian Campaign," 68. Also see Beck et al., *The Corps of Engineers,* 134. The later numbers also include use of the port of Palermo.

53. "Report of Operations, II Corps in Sicily, 10 July to 17 August, 1943," box 9, RG 338 E:P42890, NARA, 7.

54. *Report by General Dwight D. Eisenhower on Operations in the Mediterranean Area, 1942–1944,* 92.

55. *Report by General Dwight D. Eisenhower on Operations in the Mediterranean Area, 1942–1944,* 83.

56. "Report: Combined Operations Lessons Learned in the Mediterranean, 1943," box 56, RG 492, NARA, 20.

57. "History of AFHQ," vol. 1, part 1, CMH, 179.

58. Carlo D'Este, *Patton: A Genius for War* (New York: HarperCollins, 1995), 516.

59. Garland and Smyth, *Sicily and the Surrender of Italy,* 251.

60. Beck et al., *The Corps of Engineers,* 135; "Report of Operations of the United States Seventh Army in the Sicilian Campaign 10 July to 17 Aug 43," HQ Seventh Army, October 1, 1943, MHI, A-8.

61. James Dunn, "Engineers in Sicily," in *Builders and Fighters: U. S. Army Engineers in World War II,* ed. Barry Fowler (Washington, DC: US Army Corps of Engineers, 1992), 413. Also see Porch, *The Path to Victory,* 438.

62. Garland and Smyth, *Sicily and the Surrender of Italy,* 311.

63. Andrew McNamara, *Quartermaster Activities of II Corps thru Algeria, Tunisia, and Sicily and First Army thru Europe,* comp. Raymond McNally (Fort Lee, VA: US Army Publication, April 1955), 83.

64. "Report of Operations, II Corps in Sicily, 10 July to 17 August," 1.

65. Garland and Smyth, *Sicily and the Surrender of Italy,* 317.

66. "Report of Operations, II Corps in Sicily, 10 July to 17 August," 2.

67. "Report of Operations: 'Shore Engineers in Sicily,'" 1.

68. D'Este, *World War II in the Mediterranean,* 71–72; Garland and Smyth, *Sicily and the Surrender of Italy,* 329.

69. Porch, *The Path to Victory,* 439; Garland and Smyth, *Sicily and the Surrender of Italy,* 333.

70. The A-36 Apache is the ground-attack/dive-bomber version of the American P-51 Mustang. See Porch, *The Path to Victory,* 440.

71. D'Este, *Bitter Victory,* 499–500.

72. D'Este, *Bitter Victory*, 514. Also see Garland and Smyth, *Sicily and the Surrender of Italy*, 410; Rein, *The North African Air Campaign*, 135; D'Este, *World War II in the Mediterranean*, 76.

73. James Huston, *The Sinews of War: Army Logistics 1775–1953* (Washington, DC: US Army Center of Military History, 1966), 523; Lucas to Eisenhower, memorandum, August 26, 1943, subject: Sicilian Campaign, box 2775, RG 492, NARA, 3.

74. "American Rails in Eight Countries: The Story of the Military Railway Service, Army Transportation Corps," unit history, 1945, MHI, 11.

75. "American Rails in Eight Countries," 12; "Report of Operations of the United States Seventh Army in the Sicilian Campaign," B-10/11.

76. Atkinson, *The Day of Battle*, 144–46.

77. "History of the Island Base Section," unit history, [1945?], Transportation Section, CMH, 2.

78. Memorandum, October 2, 1943, subject: Administrative Lessons from Operations in Sicily, box 2775, RG 492, NARA. Also see "History of the Island Base Section," 4.

79. "History of the Island Base Section," 1.

80. General Orders Number 82, North African Theater of Operations, August 31, 1943, box 1531, RG 492, NARA. Also see "Report of Operations of the United States Seventh Army in the Sicilian Campaign," E-2; FHQ General Orders Number 62, November 5, 1943, box 1533, RG 492, NARA. Biographical information on Colonel Sears is limited.

81. "Report of Operations: 'Shore Engineers in Sicily,'" 5.

82. "Western Naval Task Force, the Sicilian Campaign," 56–57.

83. Extract from "Report on the Sicilian Campaign by the Commanding General, Seventh Army," box 3021, RG 492, NARA, 3.

84. "Notes of the Commanding General, Seventh Army, on the Sicilian Campaign," October 13, 1943, box 2775, RG 492, NARA, 6.

85. "SOS Statistical Summary," Number 4, July 1, 1943, box 2758, RG 492, NARA, 1.

86. "SOS Statistical Summary," 1.

87. "SOS Statistical Summary," 29.

88. "SOS Statistical Summary," 33.

89. Generalmajor Max Ulich and Joachim von Schoen-Angerer, "Special Experience Gained in Marches of Motorized Formations and Units in Sicily," US Historical Division MS D-063, 1947, 3–4.

90. Richard Bosworth, *Mussolini* (New York: Oxford University Press, 2002), 400.

91. Bosworth, *Mussolini*, 400.

92. Garland and Smyth, *Sicily and the Surrender of Italy*, 269.

93. Erwin Rommel, *The Rommel Papers*, ed. B. H. Liddel Hart (New York: Da Capo, 1988), 440.

94. Garland and Smyth, *Sicily and the Surrender of Italy*, 283; Porch, *The Path to Victory*, 445.

5. Operation Avalanche

1. Carlo D'Este, *World War II in the Mediterranean: 1942–1945* (Chapel Hill, NC: Algonquin, 1990), 79.

2. Winston Churchill, *Closing the Ring* (Boston: Houghton Mifflin, 1951), 112.

3. Martin Blumenson, *Salerno to Cassino* (Washington, DC: US Army Center of Military History, 2002), 11–12; Douglas Porch, *The Path to Victory: The Mediterranean Theater in World War II* (New York: Farrar, Straus and Giroux, 2004), 488–89.

4. Dwight D. Eisenhower, *Crusade in Europe* (New York: Doubleday, 1948), 199.

5. "Allied Commander-in-Chief's Report, Italian Campaign, 3 September 1943–8 January 1944," CMH, 111.

6. H. H. Dunham, "U. S. Army Transportation and the Italian Campaign," monograph prepared by the Office of Chief of Transportation, 1945, CMH, 11; Leo Meyer, "The Strategic and Logistical History of MTO," undated manuscript, CMH, xviii–23.

7. Blumenson, *Salerno to Cassino*, 67.

8. Blumenson, *Salerno to Cassino*, 22.

9. Dunham, "U. S. Army Transportation and the Italian Campaign," 20.

10. Dunham, "U. S. Army Transportation and the Italian Campaign," 24.

11. "Action Report of the Salerno Landings, the Italian Campaign Western Naval Task Force, September–October 1943," CMH, 94. Also see Dunham, "U. S. Army Transportation and the Italian Campaign," 25.

12. "Action Report of the Salerno Landings, the Italian Campaign Western Naval Task Force," 95; Rick Atkinson, *The Day of Battle: The War in Sicily and Italy, 1943–1944* (New York: Holt, 2007), 184.

13. Dunham, "U. S. Army Transportation and the Italian Campaign," 25.

14. Dunham, "U. S. Army Transportation and the Italian Campaign," 18.

15. "Action Report of the Salerno Landings, the Italian Campaign Western Naval Task Force," 104.

16. Blumenson, *Salerno to Cassino*, 46.

17. "Action Report of the Salerno Landings, the Italian Campaign Western Naval Task Force," 99.

18. "Action Report of the Salerno Landings, the Italian Campaign Western Naval Task Force," 110.

19. Howard G. DeVoe, "History of the 1st Engineer Amphibian Brigade HQ and HQ Company and the 531st Engineer Shore Regiment and Succeeding Units," unit history, 531st Engineer Association, 1983, CMH, 11.

20. DeVoe, "History of the 1st Engineer Amphibian Brigade," 11; "Report of SOS Observers of Operation Avalanche, September 9th to 21st, 1943," [1943?], box 2775, RG 492, NARA, 6.

21. D'Este, *World War II in the Mediterranean*, 93. Also see Blumenson, *Salerno to Cassino*, 97.

22. Blumenson, *Salerno to Cassino*, 98.

23. Dunham, "U. S. Army Transportation and the Italian Campaign," 36.

24. "Action Report of the Salerno Landings, the Italian Campaign Western Naval Task Force," 152.

25. "Action Report of the Salerno Landings, the Italian Campaign Western Naval Task Force," 153.

26. Joseph Sullivan, diary, entry for September 11, 1943, Joseph Sullivan Collection, US Army Quartermaster Museum, Fort Lee, VA.

27. "Action Report of the Salerno Landings, the Italian Campaign Western Naval Task Force," 153; Sullivan, diary, entry for September 17, 1943.

28. Carlo D'Este, *Fatal Decision: Anzio and the Battle for Rome* (New York: Harper-Collins, 1986), 40.

29. Porch, *The Path to Victory*, 501.

30. Blumenson, *Salerno to Cassino*, 114.

31. Dunham, "U. S. Army Transportation and the Italian Campaign," 44.

32. Porch, *The Path to Victory*, 501. Also see D'Este, *World War II in the Mediterranean*, 107.

33. Blumenson, *Salerno to Cassino*, 130.

34. Blumenson, *Salerno to Cassino*, 134.

35. "Action Report of the Salerno Landings, the Italian Campaign Western Naval Task Force," 155.

36. Dunham, "U. S. Army Transportation and the Italian Campaign," 3.

37. Dunham, "U. S. Army Transportation and the Italian Campaign," 44.

38. Dunham, "U. S. Army Transportation and the Italian Campaign," 52.

39. "Allied Commander-in-Chief's Report, Italian Campaign," 138; "History of the Peninsular Base Section, North African Theater of Operations," 3 vols., [1945?], CMH, 2:i.

40. "G4 Annex to Outline Plan—Avalanche. Headquarters, Fifth Army," August 8, 1943, Middleswart Box, Robert M. Littlejohn Collection, US Army Quartermaster Museum, 2; Dunham, "U. S. Army Transportation and the Italian Campaign," 20.

41. *History of 6th Port*, unit history booklet, Southern Line of Communication, undated, possibly 1945, MHI, 10.

42. "Action Report of the Salerno Landings, the Italian Campaign Western Naval Task Force," 156; "History of the Peninsular Base Section, North African Theater of Operations," 2:118; *History of 6th Port*, 10. Also see "History of the Peninsular Base Section, North African Theater of Operations," 2:42.

43. *History of 6th Port*, 13–14.

44. "Action Report of the Salerno Landings, the Italian Campaign Western Naval Task Force," 157; *History of 6th Port*, 14.

45. Dunham, "U. S. Army Transportation and the Italian Campaign," 52.

46. *History of 6th Port*, 17.

47. *History of 6th Port*, 14.

48. "US Forces in Austria," memorandum, November 6, 1945, subject: Questions on Italian Campaign with Answers by Field Marshal Kesselring, MHI, 3, 8–9.

49. US Army Ground Forces Observer Board, "Report of Observers in the MTO, 1942–1945," vol. 3, MHI, 13.

50. "Quartermaster Supply in the Fifth Army in World War II," manuscript, Quartermaster Section, Fifth Army, [1945?], CMH, 67.

51. "History of the Peninsular Base Section, North African Theater of Operations," 2:15; Brigadier General Ralph Tate, interview, January 19, 1949, Sidney Matthews Papers, MHI, 18.

52. "History of the Quartermaster," Peninsular BS, US Army Quartermaster Museum, 237; William F. Ross and Charles F. Romanus, *The Quartermaster Corps: Operations in the War against Germany* (Washington, DC: US Army Center of Military History, 1991), 243.

53. "Infantry Division Table of Authorization and Equipment" (1918), in *U. S. Army in the World War, 1917–1919,* 17 vols. (Washington, DC: US Army Center of Military History, 1988), 1:341.

54. "Quartermaster Supply in the Fifth Army in World War II," 70.

55. Peninsular BS, memorandum, November 14, 1943, subject: Plans for Mountain Operations, box 2775, RG 492, NARA; "Quartermaster Supply in the Fifth Army in World War II," 154.

56. Blumenson, *Salerno to Cassino,* 229.

57. "Quartermaster Supply in the Fifth Army in World War II," 71.

58. "History of the Peninsular Base Section, North African Theater of Operations," 2:101; E. R. Ellis, "Supply and Evacuation in Mountainous Operations," *Military Review* 24, no. 3 (June 1944): 74–76.

59. Ellis, "Supply and Evacuation in Mountainous Operations," 77.

60. Ernie Pyle, "The Death of Captain Waskow," wire-service article published in various papers, including *Washington Daily News,* January 10, 1944, at http://journalism.indiana.edu/resources/erniepyle/wartime-columns/the-death-of-captain-waskow/.

61. "Quartermaster Supply in the Fifth Army in World War II," 69; "History of the Peninsular Base Section, North African Theater of Operations," 2:101.

62. "Quartermaster Supply in the Fifth Army in World War II," 206. Also see "A Military Encyclopedia Based on Operations in the Italian Campaigns 1943–1945," G3 Section, 15th Army Group, [1945?], CMH, 449.

63. Lieutenant General Robert Colglazier, interviewed by William Kearney, 1984, Robert W. Colglazier Papers, MHI, 115.

64. "American Rails in Eight Countries: The Story of the Military Railway Service, Army Transportation Corps," unit history, 1945, MHI, 19; Dunham, "U. S. Army Transportation and the Italian Campaign," 69.

65. Colonel Charles D'Orsa, "The Trials and Tribulations of an Army G4," *Military Review* 25, no. 4 (July 1945): 25.

66. "American Rails in Eight Countries," 12–13.

67. "Summary of Supply Activities in the Mediterranean Theater of Operations," G3, MTOUSA, September 30, 1945, MHI, 40; "American Rails in Eight Countries," 13–14.

68. Jefferson Myers, "Military Railway Service in World War II," *Military Review* 24, no. 11 (February 1945): 36.

69. AFHQ General Order Number 60, October 22, 1943, subject: Development and Operation of All Italian Railways, box 1533, RG 492, NARA.

70. Colglazier, interview, 113.

71. *Tools of War,* pamphlet issued by the Peninsular BS, [1946?], Aberdeen Proving Ground, MD, 3.

72. *Tools of War,* 5.

73. *Tools of War,* 6, 62.

74. "History of Allied Force Headquarters (AFHQ)," part 2, sec. 1, December 1942–December 1943, CMH, 204.

75. "History of the Peninsular Base Section, North African Theater of Operations," 2:26.

76. "History of the Peninsular Base Section, North African Theater of Operations," 3:12, 2:31. Also see "Quartermaster Supply in the Fifth Army in World War II," 16.

77. Sullivan, diary, entry for January 8, 1944; "A Military Encyclopedia Based on Operations in the Italian Campaigns 1943–1945," 359.

78. Marcel Vigneras, *Rearming the French* (Washington, DC: US Army Center of Military History, 1989), 141.

79. Larkin to Eastern BS, message, December 16, 1943, box 667, RG 492, NARA; Joint Rearmament Committee, memorandum, January 15, 1944, subject: Requisitions for Maintenance of French Forces, box 667, RG 492, NARA.

80. Atkinson, *The Day of Battle,* 275–76.

81. Vigneras, *Rearming the French,* 144.

82. NATOUSA, memorandum, May 11, 1944, subject: Revision of French Rearmament Program, box 132, RG 492, NARA.

83. AFHQ, "History of Allied Force Headquarters," April 1945, MHI, 633.

84. Sullivan, diary, entry for October 8, 1943.

85. "History of the Peninsular Base Section, North African Theater of Operations," 2:59.

86. "A Military Encyclopedia Based on Operations in the Italian Campaigns 1943–1945," 519.

87. Fifth Army Surgeon, report, subject: Medical and Sanitary Data on Rome South Area of Italy, box 2737, RG 492, NARA.

88. "History of the Peninsular Base Section, North African Theater of Operations," 3:72.

89. "History, Communications Zone, NATOUSA," vol. 4, undated, probably 1945, declassified in 1948, CMH, medical chapter, 32. In April 1944, soldiers occupied 13,941 beds due to disease, 3,325 from injury, and 5,739 from battle wounds.

90. Charles Wiltse, *The Medical Department: Medical Service in the Mediterranean and Minor Theaters* (Washington, DC: US Army Center of Military History, 1965), 261.

91. Sullivan, diary, entry for January 21, 1944.

92. "A Military Encyclopedia Based on Operations in the Italian Campaigns 1943–1945," 519.

93. US Army Ground Forces Observer Board, "Report of Observers in the MTO, 1942–1945," 3:2, G4 Annex.

94. Sullivan, diary, entry for September 13, 1943.

95. "Quartermaster Supply in the Fifth Army in World War II," 113, 186.

96. Porch, *The Path to Victory*, 526.

6. Operation Shingle

1. Carlo D'Este, *Fatal Decision: Anzio and the Battle for Rome* (New York: Harper-Collins, 1986), 71.

2. Martin Blumenson, "General Lucas at Anzio," in *Command Decisions*, ed. Kent Roberts Greenfield (Washington, DC: US Army Center of Military History, 2000), 326.

3. Martin Blumenson, *Salerno to Cassino* (Washington, DC: US Army Center of Military History, 2002), 300. Churchill fell ill at the Tehran Conference and was forced to recover in North Africa prior to returning to England.

4. H. H. Dunham, "U. S. Army Transportation and the Italian Campaign," monograph prepared by the Office of Chief of Transportation, 1945, CMH, 86.

5. Rick Atkinson, *The Day of Battle: The War in Sicily and Italy, 1943–1944* (New York: Holt, 2007), 363.

6. Douglas Porch, *The Path to Victory: The Mediterranean Theater in World War II* (New York: Farrar, Straus and Giroux, 2004), 527.

7. Dunham, "U. S. Army Transportation and the Italian Campaign," 87.

8. D'Este, *Fatal Decision*, 120.

9. Carlo D'Este, *World War II in the Mediterranean: 1942–1945* (Chapel Hill, NC: Algonquin, 1990), 107; Dunham, "U. S. Army Transportation and the Italian Campaign," 92; Blumenson, *Salerno to Cassino*, 359.

10. D'Este, *Fatal Decision*, 124; Blumenson, "General Lucas at Anzio," 331. Also see D'Este, *Fatal Decision*, 402.

11. Porch, *The Path to Victory*, 528, 535; D'Este, *World War II in the Mediterranean*, 155; D'Este, *Fatal Decision*, 132.

12. Blumenson, "General Lucas at Anzio," 344.

13. D'Este, *Fatal Decision*, 96, 400.

14. Quoted in Flint Whitlock, *Desperate Valor: Triumph at Anzio* (New York: Da Capo, 2018), 341.

15. Dunham, "U. S. Army Transportation and the Italian Campaign," 94.

16. Major General John Lucas, interviewed by Sidney Matthews, May 24, 1948, box 2, Sidney Matthews Papers, MHI, 4.

17. "Supreme Allied Commanders Dispatch, Italian Campaign 8 Jan–10 May 44," MHI, 20.

18. "Supreme Allied Commanders Dispatch, Italian Campaign 8 Jan–10 May 44," 20.

19. D'Este, *World War II in the Mediterranean*, 149; "Ammo Joe at Anzio, January 22, 1944–May 31, 1944," Fifth Army publication, n.d., Aberdeen Proving Ground, MD, 3.

20. Fifth Army Advanced Command Post to Colonel Tate, memorandum, February 29, 1944, Ernest N. Harmon Papers, MHI.

21. "Ammo Joe at Anzio," 5.

22. "Ammo Joe at Anzio," 15.

23. Personal memoir, box 3a, Harmon Papers, MHI, 104–5.

24. Blumenson, *Salerno to Cassino*, 396; "US Forces in Austria," memorandum, August 25, 1946, subject: Answers by Field Marshal Kesselring to Questions Submitted by Headquarters ASFA, G-2 Section, MHI, 7.

25. Personal memoir, box 3a, Harmon Papers, MHI, 110.

26. Whitlock, *Desperate Valor*, 339, 346.

27. D'Este, *World War II in the Mediterranean*, 137.

28. General George C. Marshall, interviewed by Roy Lamsen, David Hamilton, Sidney Matthews, and Howard Smyth, July 25, 1949, box 3, Matthews Papers, MHI, 2.

29. Major General Fred Walker, Memorandum for Record, October 2, 1957, folder "Rapido River Crossing," Matthews Papers, MHI.

30. AFHQ, memorandum, November 8, 1943, subject: Operations on the Italian Mainland, box 124, RG 492, NARA.

31. "Summary of Supply Activities in the Mediterranean Theater of Operations," G3, MTOUSA, September 30, 1945, MHI, 1.

32. "Adriatic Depot History October 21, 1943 to January 1, 1944," [1944–1945?], CMH, 2–3, 15.

33. "Adriatic Depot History," 12.

34. "History of Allied Force Headquarters," vol. 3, part 1, CMH, 710.

35. "History, Communications Zone, NATOUSA," vol. 1, undated, probably 1945, declassified in 1948, CMH, 19.

36. "Report on ASF Installations in NATOSA, 25 Mar 44 by Arthur Trudeau, COL Director of Training for Army Service Forces," MHI, 3.

37. AFHQ G3 for Chief of Staff, memorandum, January 3, 1944, subject: Organization of NATOUSA, box 127, RG 492, NARA.

38. NATOUSA General Orders Number 4, February 24, 1944, box 2901, RG 492, NARA.

39. Dwight D. Eisenhower, *Crusade in Europe* (New York: Doubleday, 1948), 215.

40. D'Este, *World War II in the Mediterranean*, 189.

41. Field Marshal Sir Harold Alexander, interviewed by George Howe, part 2, 1949, box 2, Matthews Papers, MHI, 14.

42. Alexander, interview, 5, 24, 38.

43. Whitlock, *Desperate Valor*, 400–401.

44. Whitlock, *Desperate Valor*, 421.

7. Operation Dragoon

1. Jeffrey Clarke and Robert Smith, *Riviera to the Rhine: U. S. Army in World War II, the European Theater of Operations* (Washington, DC: US Army Center of Military History, 1991), 8.

2. *Report by the Supreme Allied Commander Mediterranean to the Combined Chiefs of Staff on the Operations in Southern France* (London: His Majesty's Stationery Office, 1946), CMH, 1.

3. Douglas Porch, *The Path to Victory: The Mediterranean Theater in World War II* (New York: Farrar, Straus and Giroux, 200), 664.

4. Dwight D. Eisenhower, *Crusade in Europe* (New York: Doubleday, 1948), 227; James Huston, *The Sinews of War: Army Logistics 1775–1953* (Washington, DC: US Army Center of Military History, 1966), 533.

5. Russel F. Weigley, *Eisenhower's Lieutenants: The Campaign of France and Germany 1944–1945* (Bloomington: Indiana University Press, 1981), 227.

6. Clarke and Smith, *Riviera to the Rhine,* 14–15.

7. Leo Meyer, "The Strategic and Logistical History of MTO," undated manuscript, CMH, xxiv-7.

8. Clarke and Smith, *Riviera to the Rhine,* 18.

9. Eisenhower to Wilson and the Combined Chiefs of Staff, message, June 23, 1944, in "Memorandum for Record, HQ North African Theater of Operations, US Army, G3 Section," September 12, 1944, subject: History of Planning for Operation against Southern France, MHI; *Report by the Supreme Allied Commander Mediterranean to the Combined Chiefs of Staff on the Operations in Southern France,* 12, 21.

10. *Report by the Supreme Allied Commander Mediterranean to the Combined Chiefs of Staff on the Operations in Southern France,* 23–24; Allied Force Headquarters, memorandum, July 7, 1944, subject: Operation Anvil, box 124, RG 492, NARA. Also see *The United States Army Seventh Army Report of Operations: France and Germany 1944–1945* (Nashville, TN: Battery Press, 1988), 2; Marshall to Devers and Eisenhower, message, July 16, 1944, reference no. WX66124, box 135, RG 492, NARA.

11. Eisenhower, *Crusade in Europe,* 281.

12. Meyer, "The Strategic and Logistical History of MTO," xxiv–43.

13. Eisenhower, *Crusade in Europe,* 226; Michael Howard, *The Mediterranean Strategy in the Second World War* (London: Greenhill, 1968), 60. Howard writes that considerations "were now logistics rather than strategy" (60). However, one can argue that strategy is a calculated relationship between the desired ends and the available means. As such, strategy is inherently connected with logistics.

14. Meyer, "The Strategic and Logistical History of MTO," xxv–2.

15. Milton Lehman, "Supplying Seventh Army," undated memoir, James S. Sweet Papers, MHI, 1. Also see Meyer, "The Strategic and Logistical History of MTO," xxv–7.

16. "The United States Seventh Army Report of Operations: France and Germany 1944–1945," MHI, 20.

17. Admiral H. Kent Hewitt, "The Landings of the Seventh Army in Southern France," n.d., box 4, Kent Hewitt Collection, US Navy Archives, Navy Yard, Washington, DC, 9.

18. *Report by the Supreme Allied Commander Mediterranean to the Combined Chiefs of Staff on the Operations in Southern France,* 25.

19. "Final Report: G-3 Section, Headquarters, 6th Army Group, 1945," CMH, 5.

20. Eisenhower to General George Marshall, message, December 1943, John P. Ratay Photograph Collection, 1908–1945, MHI; George C. Marshall, *The Papers of George Catlett Marshall*, vol. 4: *Aggressive and Determined Leadership, June 1, 1943–December 31, 1944*, ed. Larry I. Bland and Sharon Ritenour Stevens (Baltimore: Johns Hopkins University Press, 1996), 83–84.

21. *CONAD History, Continental Advance Section, Communications Zone, European Theater of Operations* (Heidelberg, Germany: Aloys Graff, 1945), 5.

22. Clarke and Smith, *Riviera to the Rhine*, 355; Meyer, "The Strategic and Logistical History of MTO," xxv–15, 49. Also see "The United States Seventh Army Report of Operations: France and Germany 1944–1945," 67.

23. "The United States Seventh Army Report of Operations: France and Germany 1944–1945," 67.

24. Meyer, "The Strategic and Logistical History of MTO," xxv–36.

25. James L. Whelchel, "Activities of the Quartermaster in the Continental Advance Section, European Theater of Operation," in US Army QM Center, *Passing in Review*, ed. Robert Littlejohn (Fort Lee, VA: US Army Quartermaster Center, 1955), chap. 43, pp. 3–4 (each chapter paginated separately).

26. "The United States Seventh Army Report of Operations: France and Germany 1944–1945," 65.

27. *Report by the Supreme Allied Commander Mediterranean to the Combined Chiefs of Staff on the Operations in Southern France*, 6.

28. Meyer, "The Strategic and Logistical History of MTO," xxvii–2.

29. *Report by the Supreme Allied Commander Mediterranean to the Combined Chiefs of Staff on the Operations in Southern France*, 6.

30. "Report: Invasion of Southern France. Report of Naval Commander, Western Task Force," August 15, 1944, box 4, Hewitt Collection, US Navy Archives, 169.

31. "Report: Invasion of Southern France. Report of Naval Commander, Western Task Force," 324.

32. Lehman, "Supplying Seventh Army," 1.

33. Lehman, "Supplying Seventh Army," 1.

34. Lehman, "Supplying Seventh Army," 34.

35. *Report by the Supreme Allied Commander Mediterranean to the Combined Chiefs of Staff on the Operations in Southern France*, 33.

36. Hewitt, "The Landings of the Seventh Army in Southern France," 16.

37. Clarke and Smith, *Riviera to the Rhine*, 122–23.

38. *The United States Army Seventh Army Report of Operations*, 316.

39. *The United States Army Seventh Army Report of Operations*, 162.

40. *The United States Army Seventh Army Report of Operations*, 323, 318; Meyer, "The Strategic and Logistical History of MTO," xxvi–22.

41. Meyer, "The Strategic and Logistical History of MTO," xxvii–6; "From the Sahara to the Rhine, a History of Army Supply Service," Southern Line of Communication Publication, Communications Zone ETOUSA, 1945, MHI, 14.

42. *Report by the Supreme Allied Commander Mediterranean to the Combined Chiefs of Staff on the Operations in Southern France,* 39; *History of 6th Port,* unit history booklet, Southern Line of Communication, undated, possibly 1945, MHI, 25; "The Administrative and Logistical History of the European Theater of Operations, Part VII: Opening and Operating Continental Ports," CMH, figure 1, p. 12.

43. Eisenhower, *Crusade in Europe,* 260, 280, 290.

44. *History of 6th Port,* 25.

45. "Report: Invasion of Southern France. Report of Naval Commander, Western Task Force," 230.

46. "From the Sahara to the Rhine," 15.

47. *CONAD History,* 47; Clarke and Smith, *Riviera to the Rhine,* 200.

48. *Report by the Supreme Allied Commander Mediterranean to the Combined Chiefs of Staff on the Operations in Southern France,* 44.

49. *CONAD History,* 69.

50. US Army Ground Forces Observers Board, "Report of Observers in the MTO, 1942–1945," vol. 6, March 21 to October 22, 1944, MHI, 22.

51. US Army Ground Forces Observers Board, "Report of Observers in the MTO, 1942–1945," 6:1; "Army Ground Forces Report Number 232," December 4, 1944, in US Army Ground Forces Observers Board, "Report of Observers in the MTO, 1942–1945," 6:5.

52. Lehman, "Supplying Seventh Army," 2.

53. Lehman, "Supplying Seventh Army," 2.

54. "American Rails in Eight Countries: The Story of the Military Railway Service, Army Transportation Corps," unit history, 1945, MHI, 22.

55. "American Rails in Eight Countries," 23.

56. Meyer, "The Strategic and Logistical History of MTO," xxvii–20.

57. Whelchel, "Activities of the Quartermaster in the Continental Advance Section,"13.

58. "American Rails in Eight Countries," 24; Meyer, "The Strategic and Logistical History of MTO," xxvii–25.

59. *CONAD History,* 234.

60. *CONAD History,* 236–38.

61. *CONAD History,* 237; "Report: The 6th Army Group, France and Germany, 1944–1945, with Special Attention to Logistical Problems, G4 Section 6th Army Group," n.d., box 1, Frank Osmanski Papers, MHI, 10.

62. Marcel Vigneras, *Rearming the French* (Washington, DC: US Army Center of Military History, 1989), 259.

63. Vigneras, *Rearming the French,* 320.

64. Vigneras, *Rearming the French,* 337.

65. *The United States Army Seventh Army Report of Operations,* 70.

66. Clarke and Smith, *Riviera to the Rhine,* 292.

67. *The United States Army Seventh Army Report of Operations,* 531

68. "Final Report: G-3 Section, Headquarters, 6th Army Group, 1945," CMH, 15.

69. Clarke and Smith, *Riviera to the Rhine,* 439; Rick Atkinson, *The Guns at Last Light: The War in Western Europe 1944–1945* (New York: Holt, 2013), 375.

70. David Colley, *Decision at Strasbourg: Ike's Strategic Mistake to Halt the Sixth Army Group at the Rhine in 1944* (Annapolis, MD: Naval Institute Press, 2008), 142, 170; Clarke and Smith, *Riviera to the Rhine,* 444.

71. "Report of Operations, 12th Army Group, 31 July 1945," vol. 1, MHI, 7, 25.

72. Clarke and Smith, *Riviera to the Rhine,* 490.

73. Julian Thompson, *The Lifeblood of War: Logistics in Armed Conflict* (New York: Brassey's, 1991), 12.

74. A K ration is an individual combat ration (Thompson, *The Lifeblood of War,* 5). Fuel levels increased every winter due to soldiers' use of heaters and running vehicles to keep themselves warm.

75. Clarke and Smith, *Riviera to the Rhine,* 485.

76. Colley, *Decision at Strasbourg,* 185.

77. Clarke and Smith, *Riviera to the Rhine,* 502, 526.

78. "Report: The 6th Army Group, France and Germany," 10.

79. General Orders Number 38, North African Theater of Operations, September 11, 1944, box 2901, RG 492, NARA; *CONAD History,* 70.

80. COMZONE Statistical Study Number 19, October 1, 1944, CMH.

81. *CONAD History,* 72. Also see General Orders Number 42, North African Theater of Operations, September 26, 1944, box 2901, RG 492, NARA.

82. North African Theater of Operations, memorandum, October 6, 1944, subject: Assignment of Units, box 3020, RG 492, NARA.

83. Vigneras, *Rearming the French,* 187; *CONAD History,* 82–83.

84. General Orders Number 24, Allied Force Headquarters, October 22, 1944, box 1533, RG 492, NARA.

85. Colley, *Decision at Strasbourg,* 15.

86. *CONAD History,* 112.

87. Whelchel, "Activities of the Quartermaster in the Continental Advance Section," 25.

88. Meyer, "The Strategic and Logistical History of MTO," xxviii–38; "From the Sahara to the Rhine," 14.

89. General George C. Marshall, interviewed by Sidney Matthews and Howard Smyth, July 25, 1949, box 3, Sidney Matthews Papers, MHI, 10.

90. Eisenhower, *Crusade in Europe,* 243.

91. "Report: Invasion of Southern France. Report of Naval Commander, Western Task Force," 337.

92. "6th Army Group History," vol. 3, 1945, MHI, 370.

93. "Report: The 6th Army Group, France and Germany," 12.

94. "Report: The 6th Army Group, France and Germany," 129.

95. "Report: The 6th Army Group, France and Germany," 134.

96. "Report: The 6th Army Group, France and Germany," 143–44.

8. Unfinished Business

1. Joseph Bykofsky and Harold Larson, *The Transportation Corps: Operations Overseas* (Washington, DC: US Army Center of Military History, 2003), 212.

2. Clark to Wilson, report, June 27, 1944, box 135, RG 492, NARA; Bykofsky and Larson, *The Transportation Corps*, 212.

3. "Tools of War: An Illustrated History of the Peninsular Base Section," unit history, 1946, Aberdeen Proving Ground, MD, 16.

4. Bykofsky and Larson, *The Transportation Corps*, 212, 214.

5. Joseph Sullivan, "History of Fifth Army Quartermaster Activities," undated manuscript, US Army Quartermaster Museum, Fort Lee, VA, 125.

6. Sullivan, "History of Fifth Army Quartermaster Activities," 132.

7. G4, 1st Armored Division, to Commanding General, 1st Armored Division, memorandum, July 4, 1944, subject: Status of Vehicles, Ernest Harmon Papers, MHI.

8. Ernest Fisher, *Cassino to the Alps* (Washington, DC: US Army Center of Military History, 2002), 337.

9. *Administrative Planning* (London: British War Office, 1952), 71–72.

10. Oral history, Carter Magruder Papers, MHI, 45–46.

11. Fisher, *Cassino to the Alps*, 447, 458.

12. Fisher, *Cassino to the Alps*, 456.

13. Oral history, Charles Bolte Papers, sec. 3, MHI, 6.

14. "Tools of War," 2–8.

15. Joseph Sullivan, diary, entry for December 8, 1944, Joseph Sullivan Collection, US Army Quartermaster Museum.

16. Special Text 35–150, *Role of the WAC* (Fort McLellan, AL: United States Women's Army Corps School, 1962), preface.

17. Mattie Treadwell, *The Women's Army Corps* (Washington, DC: US Army Center of Military History, 1991), 360; Judith Bellafaire, *The Women's Army Corps: A Commemoration of World War II Service* (Washington, DC: US Army Center of Military History, 2005), at https://history.army.mil/brochures/WAC/WAC.HTM.

18. Special Text 35–150, *Role of the WAC*, 27; Colonel Bettie Morden, "Women's Army Corps: WAAC and WAC," in *In Defense of a Nation: Servicewomen in World War II*, ed. Jeanne M. Holm (Washington, DC: Military Women's Press, 1998), 45–46; Treadwell, *The Women's Army Corps*, 361.

19. Treadwell, *The Women's Army Corps*, 265, 368.

20. Special Text 35–150, *Role of the WAC*, 28; Bellafaire, *The Women's Army Corps*.

21. *Awards and Citations to Women in the Armed Services (during WWII)* (Washington, DC: Office of War Information, 1944), MHI, 1–4.

22. *Study of the Women's Army Corps in the European Theater of Operations*, vol. 2 (Washington, DC: General Board, United States Forces, European Theater, 1945), chap. 4, p. 1 (each chapter paginated separately).

23. H. R. Bull, Major General, "Statement Regarding the W.A.C.," n.d., MHI.

24. A. Russel Buchanan, *Black Americans in World War II* (Santa Clara, CA: Clio Books, 1977), 1–2; Ulysses Lee, *The Employment of Negro Troops* (Washington, DC: US Army Center of Military History, 1966), 5.

25. Buchanan, *Black Americans in World War II*, 64; Horace Mann Bond, "The Negro in the Armed Forces of the United States prior to World War I," in "The American Negro in World Wars I and II," *Journal of Negro Education* Yearbook 2 (Summer 1943): 287.

26. Lee, *The Employment of Negro Troops*, 406.

27. Bryan D. Booker, *African Americans in the United States Army in World War II* (Jefferson, NC: McFarland, 1956), 61. Also see Lee, *The Employment of Negro Troops*, 128–29, 623.

28. Booker, *African Americans in the United States Army in World War II*, 86. Also see Lee, *The Employment of Negro Troops*, 634–36.

29. Booker, *African Americans in the United States Army in World War II*, 86.

30. Russel F. Weigley, *Eisenhower's Lieutenants: The Campaign of France and Germany 1944–1945* (Bloomington: Indiana University Press, 1981), 271; John Silvera, *The Negro in World War II* (New York: Arno Press, 1969), appendix 3.

9. Impact and Conclusion

1. Sir Frederick Morgan, *Peace and War: A Soldier's Life* (London: Hodder and Stoughton, 1961), 150–51.

2. Lieutenant Clifford Jones, "The Administrative and Logistical History of the ETO," 11 parts, Historical Services Division, US European Theater of Operations, MHI.

3. Jones, "The Administrative and Logistical History of the ETO," part 2, [no vol. no.], 164.

4. Jones, "The Administrative and Logistical History of the ETO," part 4, vol. 1, 21.

5. Roland Ruppenthal, *The European Theater of Operations: Logistical Support of the Armies,* vol. 1: *May 1941–September 1944* (Washington, DC: US Army Center of Military History, 1995), 331; Alfred Beck, Abe Bortz, Charles Lynch, Lida Mayo, and Ralph Weld, *The Corps of Engineers: The War against Germany* (Washington, DC: US Army Center of Military History, 1998), 121; Jones, "The Administrative and Logistical History of the ETO," part 4, vol. 1, 22; Douglas Porch, *The Path to Victory: The Mediterranean Theater in World War II* (New York: Farrar, Straus and Giroux, 2004), 412 (quote from Bradley).

6. Carlo D'Este, *Decision in Normandy* (Old Saybrook, CT: Konecky & Konecky, 1994), 39–40; Christopher Rein, *The North African Air Campaign: U. S. Army Air Forces from El Alamein to Salerno* (Lawrence: University Press of Kansas, 2012), 149.

7. Ruppenthal, *The European Theater of Operations,* 1:332.

8. Jones, "The Administrative and Logistical History of the ETO," part 4, vol. 1, 40, 53–57; Ruppenthal, *The European Theater of Operations,* 1:334; D'Este, *Decision in Normandy,* 57.

9. Jones, "The Administrative and Logistical History of the ETO," part 4, vol. 1, 40–42; Rick Atkinson, *The Day of Battle: The War in Sicily and Italy, 1943–1944* (New

York: Holt, 2007), 55. Also see Carlo D'Este, *World War II in the Mediterranean: 1942–1945* (Chapel Hill, NC: Algonquin, 1990), 74.

10. Quoted in Carlo D'Este, *Patton: A Genius for War* (New York: HarperCollins, 1995), 514.

11. D'Este, *Decision in Normandy,* 39–40; Dwight D. Eisenhower, *Crusade in Europe* (New York: Doubleday, 1948), 222.

12. Jones, "The Administrative and Logistical History of the ETO," part 4, vol. 1, 49; Ruppenthal, *The European Theater of Operations,* 1:334.

13. Jones, "The Administrative and Logistical History of the ETO," part 4, vol. 1, 51; Rick Atkinson, *An Army at Dawn: The War in North Africa, 1942–1943* (New York: Holt, 2002), 540.

14. "Royal Navy (RN) Officers 1939–1945," n.d., at https://www.unithistories.com/officers/RN_officersV.html.

15. Historical Sub-section, Office of Secretary, General Staff, Supreme Headquarters, Allied Expeditionary Force, "The Development of COSSAC," in *History of COSSAC (Chief of Staff to Supreme Allied Commander),* ed. Ray Merriam (Bennington, VT: Merriam Press, 2000), 22; Forrest C. Pogue, *The Supreme Command* (Washington, DC: US Army Center of Military History, 1996), 56.

16. D. K. R. Crosswell, *Beetle: The Life of General Walter Bedell Smith* (Lexington: University Press of Kentucky, 2010), 554–55.

17. Pogue, *The Supreme Command,* 64.

18. Porch, *The Path to Victory,* 530.

19. Historical Sub-section, Office of Secretary, General Staff, Supreme Headquarters, Allied Expeditionary Force, "The Development of COSSAC," 19; "Royal Navy (RN) Officers 1939–1945."

20. Carlo D'Este, *Bitter Victory: The Battle for Sicily, 1943* (New York: Dutton, 1988), 84.

21. Hank H. Cox, *The General Who Wore Six Stars* (Lincoln: Potomac Books, University of Nebraska Press, 2018), 62–64, 69.

22. Cox, *The General Who Wore Six Stars,* 80.

23. Steve R. Waddell, *United States Army Logistics: The Normandy Campaign, 1944* (Westport, CT: Greenwood Press, 1994), 3.

24. Waddell, *United States Army Logistics,* 19.

25. Crosswell, *Beetle,* 615.

26. Crosswell, *Beetle,* 842–45.

27. Albert Kesselring, "Final Commentaries on the Campaign in North Africa, 1941–1954," US Historical Division MS C-075, 1949, MHI, 11–12, 24.

28. Kesselring, "Final Commentaries on the Campaign in North Africa, 1941–1954," 44.

29. Hans von Arnim, "Recollections of Tunisia," US Historical Division MS C-098, 1951, MHI, 3, 7–8, 10.

30. Major Richard Feige, "The Relationship between Operations and Supply in Africa," MS D-125, 1947, MHI, 2–4, 12.

31. Porch, *The Path to Victory,* 664–69.

32. *American Enterprise in Europe: The Role of the Services of Supply in the Defeat of Germany,* unit historical pamphlet, n.d., US Army Quartermaster Museum, Fort Lee, VA, 16–17.

33. *American Enterprise in Europe,* 116–21.

34. "Summary of Supply Activities in the Mediterranean Theater of Operations," MHI, 2.

35. The information on US, German, and Russian forces was provided by Dr. John Bonin, US Army War College, from research conducted on average divisional slice compositions as of June 1944. The data on British forces are based on planning factors listed in *Administrative Planning* (London: British War Office, 1952), 182, and reflects a planning figure rather than an average strength for any specific period.

36. Eisenhower, *Crusade in Europe,* 234.

Index